"Martin Gardner's contribution to contemporary intellectual culture is unique—in its range, its insight, and its understanding of hard questions that matter." —Noam Chomsky

"Nobody has worked harder or more steadily to defend and enlarge this little firelit clearing we hold in the dark chittering forest of unreason." —*The New Criterion*

"Martin Gardner is a national treasure. . . . [This book] should be compulsory reading in every high school and in Congress." —Arthur C. Clarke

"What sets Gardner apart from the rest of the debunking crowd is that he doesn't just take potshots at the sitting ducks of fundamentalist thought or kooky cults; the sacred cows of science appear equally—and hilariously—in his deadly cross hairs." —*Salon.com*

"A recreational excursion among ignoramuses." —*Booklist*

D0187993

Books by Martin Gardner

Fads and Fallacies in the Name of Science
Mathematics, Magic, and Mystery
Great Essays in Science (ed.)
Logic Machines and Diagrams
The Scientific American Book of Mathematical Puzzles and Diversions
The Annotated Alice
The Second Scientific American Book of Mathematical Puzzles and Diversions
Relativity for the Million
The Annotated Snark
The Ambidextrous Universe
The Annotated Ancient Mariner
New Mathematical Diversions from Scientific American
The Annotated Casey at the Bat
Perplexing Puzzles and Tantalizing Teasers
The Unexpected Hanging and Other Mathematical Diversions
Never Make Fun of a Turtle, My Son (verse)
The Sixth Book of Mathematical Games from Scientific American
Codes, Ciphers, and Secret Writing
Space Puzzles
The Snark Puzzle Book
The Flight of Peter Fromm (novel)
Mathematical Magic Show
More Perplexing Puzzles and Tantalizing Teasers
The Encyclopedia of Impromptu Magic
Aha! Insight
Mathematical Carnival
Science: Good, Bad, and Bogus
Science Fiction Puzzle Tales
Aha! Gotcha
Wheels, Life, and Other Mathematical Amusements
Order and Surprise
The Whys of a Philosophical Scrivener
Puzzles from Other Worlds
The Magic Numbers of Dr. Matrix
Knotted Doughnuts and Other Mathematical Entertainments

The Wreck of the Titanic Foretold?
Riddles of the Sphinx
The Annotated Innocence of Father Brown
The No-Sided Professor (short stories)
Time Travel and Other Mathematical Bewilderments
The New Age: Notes of a Fringe Watcher
Gardner's Whys and Wherefores
Penrose Tiles to Trapdoor Ciphers
How Not to Test a Psychic
The New Ambidextrous Universe
More Annotated Alice
The Annotated Night Before Christmas
Best Remembered Poems (ed.)
Fractal Music, Hypercards, and More
The Healing Revelations of Mary Baker Eddy
Martin Gardner Presents
My Best Mathematical and Logic Puzzles
Classic Brainteasers
Famous Poems of Bygone Days (ed.)
Urantia: The Great Cult Mystery
The Universe Inside a Handkerchief
On the Wild Side
Weird Water and Fuzzy Logic
The Night Is Large
Mental Magic
The Annotated Thursday
Visitors from Oz (novel)
The Annotated Alice: The Definitive Edition

Did Adam and Eve Have Navels?

Debunking Pseudoscience

Martin Gardner

W. W. Norton & Company   New York   London

First published as a Norton paperback 2001

Excerpt from Thomas Nagel's review on pages 151–52
reprinted with permission of *The New Republic*.

For information about permission to reproduce selections from this book, write to
Permissions, W. W. Norton & Company, Inc. 500 Fifth Avenue, New York, NY 10110

The text of this book is composed in Adobe Garamond
with the display set in Meta
Composition by Allentown Digital Services
Division of R.R. Donnelley & Sons Company
Manufacturing by Haddon Craftsmen, Inc.
Book design by Chris Welch

Library of Congress Cataloging-in-Publication Data

Gardner, Martin, 1914–
Did Adam & Eve have navels? : discourses on reflexology, numerology, urine therapy &
other dubious subjects / Martin Gardner.
p.   cm.
Includes bibliographical references and index.
ISBN 0-393-04963-9
1. Science—Miscellanea. I. Title: Did Adam and Eve have navels?. II. Title.

Q173 .G34 2000
500—dc21
00-034870

ISBN 0-393-32238-6 pbk.

W. W. Norton & Company, Inc. 500 Fifth Avenue, New York, N.Y. 10110
www.wwnorton.com

W. W. Norton & Company Ltd., Castle House, 75/76 Wells Street, London W1T 3QT

4 5 6 7 8 9 0

To Kendrick Frazier, editor of the *Skeptical Inquirer*, friend, and leader in the never-ceasing battle against superstition, paranormal nonsense, and dubious science.

Contents

Introduction

Most of the chapters in this collection are attacks on far-out cases of pseudoscience. I am aware of the difficulties involving what philosophers of science call the "demarcation problem"—the task of formulating sharp criteria for distinguishing good science from bad. Clearly no such criteria are precise. Pseudoscience is a fuzzy word that refers to a vague portion of a continuum on which there are no sharp boundaries.

At the far left end of this spectrum are beliefs which all scientists consider preposterous. Examples include claims that the earth is a hollow sphere and we occupy its interior, that the earth was created in six literal days about ten thousand years ago, and that positions of stars correlate with character and future events. Moving to the right, toward slightly less weird claims, we come upon Velikovsky's cosmology, homeopathy, phrenology, Scientology, the orgone theories of Wilhelm Reich, and scores of other bizarre medical and psychiatric fads.

As we move along the continuum toward more respectable science, we reach such controversial claims as the conjectures of Freud, the belief that God pushed evolution along with little miracles, efforts to extract unlimited energy from the vacuum of space, Hans Arp's attack on the red shift and his claim that quasars are nearby objects, and a raft of other speculations in areas where there is some evidence, but much greater doubt.

At the far right end, our spectrum fades into regions of open conjectures by scientists so eminent that no one dares call them cranks. I am thinking of David Bohm's pilot-wave theory of quantum mechanics, Roger Penrose's twistors, superstrings, speculations about a myriad of other universes, the notion that life came here from outer space, and ongoing efforts by physicists to construct a TOE (theory of everything). To the right of these reputable conjectures lie the undisputed facts of science, such as that galaxies contain billions of stars, that water boils and freezes, and that dinosaurs once roamed the earth; there are millions of such claims that no informed person of sound mind doubts.

All the chapters in this anthology except one are reprints of my column "Notes of a Fringe Watcher," which appears regularly in *Skeptical Inquirer*. This lively bimonthly, so ably edited by Kendrick Frazier, is the official organ of CSICOP, the Committee for the Scientific Investigation of Claims of the Paranormal. The exception is the chapter on the Wandering Jew, which was an article in *Free Inquiry*.

Although "debunker" is often considered a pejorative term, I do not find it so. A major purpose of *Skeptical Inquirer* has always been to debunk the more outrageous claims of bogus science. I make no apologies for being a debunker. I believe it is the duty of both scientists and science writers to keep exposing the errors of bad science, especially in medical fields, in which false beliefs can cause needless suffering and even death.

We know from polls how ignorant the general public is about science. Almost half of all adults in the United States now believe in astrology and in angels and demons, and that we are being observed by aliens in UFOs who frequently abduct humans. More than half believe that evolution is an unverified theory.

Science education in our nation, especially in lower grades, is getting worse, not better. Several states are constantly doing their best to force public schools to teach creationism. Greedy publishers, interested only in profit, turn out book after book on astrology, ufology, the occult, dangerous programs to lose weight without exercising or cutting calories, and every known variety of dubious medicine.

The electronic media are equal offenders. Every year I hope the tide is about to turn, and that contributors to television, radio, and the Internet will become so appalled by the flood of fake science they keep flinging at the public that they will at least try to tone it down. Alas, every year the flood gets worse. As for book publishers, to be impressed by the flood's magnitude you need only visit any mall bookstore and compare the size of its New Age or metaphysical section with its science section. Books on astrology far outnumber books on astronomy. As the late Carl Sagan liked to point out, there are more professional astrologers in the United States than there are astronomers. The scene is just as dismal, if not worse, in other countries.

I'm not sure why I became interested in debunking bad science. It may have been my disenchantment with the views of George McCready Price. Price was an uneducated Seventh-day Adventist whose many books defending a six-day creation and the flood theory of fossils I took seriously for a very brief period in my boyhood. It was not until I attended classes in biology and geology at the University of Chicago that I finally understood where Price went wrong and what an amusing dunce he was.

At any rate, after I found that the evidence for evolution was as overwhelming as the "theory" that the earth goes around the sun—when theories become strongly confirmed they become "facts"—I wrote an article titled "The Hermit Scientist" that appeared in the *Antioch Review.* A high-school friend who had become a Manhattan literary agent persuaded me to expand that article into a book, which he subsequently placed with Putnam. Titled *In the Name of Science,* it was quickly remaindered, but Dover picked it up, and as a paperback it became one of Dover's early best-sellers.

Its sales were largely due to continual attacks on it by guests on the all-night radio show of Long John Nebel, the precursor of Art Bell. Bell's radio show, like Nebel's, owes its popularity to interviewing crackpots.

My book on pseudoscience led philosopher Paul Kurtz to contact me, along with magician James Randi, psychologist Ray Hyman, and sociologist Marcello Truzzi, to organize the group that became CSICOP in 1976. I have many other interests more important than pseudoscience, but the topic has provided material for four anthologies: *Science: Good, Bad, and Bogus; The New Age; On the Wild Side;* and *Weird Water and Fuzzy Logic.* This is the fifth such collection. I don't expect any of those books, including this one, to alter minds set in concrete, but if occasionally they help an open-minded reader to discard a crazy belief, they may do more than simply provide entertainment and laughter for skeptics.

Martin Gardner
Hendersonville, N.C.

Note added to this edition:

On page 112 I mention that no reader has identified the source of Freud's alleged remark about a cigar. In 2002 I received a letter from Richard Fischer about his failed effort to track down the quote. The anecdote is recounted in *The Little, Brown Book of Anecdotes*, edited by Clifton Fadiman (1985), on page 223. The source of the quote is given as the *Los Angeles Times*, July 4, 1982. No local library had access to the *Times* before 1990, nor (surprisingly) did the *Los Angeles Times*.

"Not to be discouraged," Fischer writes, "and living in the Washington, D.C., area, I headed off to the Library of Congress to find their July 1982 *Los Angeles Times* microfiche 'missing' and 'may not be replaced.'"

So, the mystery remains. I would be grateful if any reader could track down what the *Los Angeles Times* has to say about the source for Freud's supposed remark.

Part I

Evolution

vs.

Creationism

CHAPTER I

Did Adam and Eve Have Navels?

What did Adam and Eve never have,
yet they gave two of them to each
of their children?
Answer: Parents.
                              —*Old children's riddle*

I f you ever find yourself in the company of a fundamentalist, much pleasant argumentation can result if you ask him or her a simple question: Did Adam and Eve have belly buttons?

For those who believe the Bible to be historically accurate, this is not a trivial question. If Adam and Eve did *not* have navels, then they were not perfect human beings. On the other hand, if they *had* navels, then the navels would imply a birth they never experienced.

Bruce Felton and Mark Fowler are the authors of *The Best, Worst, and Most Unusual* (Galahad Books, 1994). In this entertaining reference work, they devote several paragraphs (pp. 146–47) to what they call "the worst theological dispute." They take this to be the acrimonious debate, which has been going on ever since the Book of Genesis was written, over whether the first human pair had what Sir Thomas Browne, in 1646, referred to as "that tortuosity or complicated nodosity we usually call the Navell."

Browne's opinion was that Adam and Eve, because they had no parents, must have had perfectly smooth abdomens. In 1752, according to Felton and Fowler, the definitive treatise on the topic was published in Germany. It was titled *Untersuchung der Frage: Ob unsere ersten Uraltern, Adam and Eve, einen Nabel gehabt (Examination on the Question: Whether Our First Ancestors, Adam and Eve, Possessed a Navel)*. After discussing all sides of this difficult question, the author, Dr. Christian Tobias Ephraim Reinhard, finally concluded that the famous pair were navelless.

As Felton and Fowler tell us, some paintings of Adam and Eve from the Middle Ages and early Renaissance show navels; others do not. Michelangelo's Sistine Chapel painting of Adam being created by God's finger shows Adam with a navel. Most artists of later periods followed Michelangelo's lead.[1]

In 1944 the old conundrum had a hilarious revival in the United States Congress. A Public Affairs booklet titled "The Races of Mankind," by Columbia University anthropologists Ruth Benedict and Gene Weltfish, was amusingly illustrated by Ad Reinhardt. Reinhardt later became notorious as an abstract expressionist who painted canvases that were solid black, or blue, or some other single color. One of his cartoons in the Public Affairs Pamphlet No. 85 had a little black dot on the abdomens of Adam and Eve.

Congressman Carl T. Durham of North Carolina and his House Military Affairs Committee were not amused. They believed that distribution of the government pamphlet to American servicemen would be an insult to those who were fundamentalists. As Felton and Fowler point out, some cynics suspected that the congressmen really objected to a table in the booklet showing that northern blacks scored higher on Army Air Force in-

---

[1]How British artists handled the navel problem is discussed along with many reproductions of paintings in Horace Walpole's four-volume *Anecdotes of Paintings,* vol. 1, ch. 3. This monumental work, published in England during the years 1762 to 1771, was later expanded and revised by other authors. It may still be in print as an Ayer reprint.

telligence tests than southern whites. I suspect that another basis for their opposition to the booklet was their belief that Weltfish was a Communist, based on her refusal to testify whether she was or was not a member of the Communist Party. Years later, in 1953, she was much in the news when she charged the United States with using germ warfare in Korea.

The old question about the navels of Adam and Eve figured prominently in one of the strangest books ever written. The book, written by an eminent scientist who wished to defend the accuracy of Genesis, was titled *Omphalos: An Attempt to Untie the Geological Knot,* and it was published in England in 1857, two years before Darwin's *Origin of Species.*

Omphalos is the Greek word for *navel.* A wonderful ancient myth tells how Zeus, in an effort to determine the exact center of a circular flat earth, had two eagles fly at the same speed from opposite ends of one of the circle's diameters. They met at Delphi. To mark the spot, a piece of white marble, called the Omphalos Stone, was placed in Apollo's temple at Delphi with a gold eagle on each side of the stone. The stone was often depicted on Greek coins and vases, usually in the shape of half an egg. (See William J. Woodhouse's detailed article "Omphalos" in James Hastings' *Encyclopaedia of Religion and Ethics.*)

The author of *Omphalos* was British zoologist Philip Henry Gosse (1810–88), father of Sir William Edmund Gosse (1849–1928), a noted English poet and critic.[2] A fundamentalist in the Plymouth Brethren sect, the elder Gosse realized that fossils of plants and animals strongly implied life that predated Adam and Eve. At the same time, he was certain that the entire universe was created in six literal days about four thousand years before Christ.

Was there any way to harmonize this stark contradiction between Genesis and the fossil record? Gosse was struck by what Jorge Luis Borges

---

[2]Philip Gosse's many books include *Canadian Naturalist; Introduction to Zoology; The Ocean; A Naturalist's Rambles on the Devonshire Coast; Acquarium; Birds of Jamaica; Naturalist's Sojourn in Jamaica; A Manual of Marine Zoology* (two volumes); *Life; Actinologia Britannica; The Romance of Natural History;* and other popular books. The eleventh edition of the *Encyclopaedia Britannica* says that for a time he taught zoology in Alabama.

would later call an idea of "monstrous elegance." If God created Adam and Eve with navels, implying a birth they never had, could not God just as easily have created a record of a past history of the earth that never existed except in the Divine Mind?

As Gosse realized, it is not just a question of belly buttons. Adam and Eve had bones, teeth, hair, fingernails, and all sorts of other features that contained evidence of previous growth. Allow me to quote at length from my 1952 book *Fads and Fallacies in the Name of Science:*

> The same is true of every plant and animal. As Gosse points out, the tusks of an elephant exhibit past stages, the nautilus keeps adding chambers to its shell, the turtle adds laminae to its plates, trees bear the annual rings of growth produced by seasonal variations. "Every argument," he writes, "by which the physiologist can prove . . . that yonder cow was once a foetus . . . will apply with exactly the same power to show that the newly created cow was an embryo some years before creation." All this is developed by the author in learned detail, for several hundred pages, and illustrated with dozens of wood engravings.
>
> In short—if God created the earth as described in the Bible, he must have created it a "going concern." Once this is seen as inevitable, there is little difficulty in extending the concept to the earth's geologic history. Evidence of the slow erosion of land by rivers, of the twisting and tilting of strata, mountains of limestone formed by remains of marine life, lava which flowed from long-extinct volcanoes, glacier scratchings upon rock, footprints of prehistoric animals, teeth marks on buried bones, and millions of fossils sprinkled through the earth—all these and many other features testify to past geological events which *never actually took place.*
>
> "It may be objected," writes Gosse, "that to assume the world to have been created with fossil skeletons in its crust—skeletons of animals that never really existed—is to charge the Creator with forming objects whose sole purpose was to deceive us. The reply is obvious. Were the concentric timber-rings of a created tree formed merely to deceive? Were the growth lines of a created shell intended to deceive? Was the navel of the created Man intended to deceive him into the persuasion that he had a parent?"
>
> So thorough is Gosse in covering every aspect of this question that he

even discusses the finding of coprolites, fossil excrement. Up until now, he writes, this "has been considered a more than ordinarily triumphant proof of real pre-existence." Yet, he points out, it offers no more difficulty than the fact that waste matter would certainly exist in the intestines of the newly formed Adam. Blood must have flowed through his arteries, and blood presupposes chyle and chyme, which in turn presupposes an indigestible residuum in the intestines. "It may seem at first sight ridiculous," he confesses, ". . . but truth is truth."

Gosse's argument is, in fact, quite flawless. Not a single truth of geology need be abandoned, yet the harmony with Genesis is complete. As Gosse pointed out, we might even suppose that God created the earth a few minutes ago, complete with all its cities and records, and memories in the minds of men, and there is no logical way to refute this as a possible theory.

Nevertheless, *Omphalos* was not well received. "Never was a book cast upon the waters with greater anticipation of success than was this curious, this obstinate, this fanatical volume," writes the younger Gosse in his book *Father and Son*. ". . . He offered it, with a glowing gesture, to atheists and Christians alike. . . . But, alas! atheists and Christians alike looked at it and laughed, and threw it away . . . even Charles Kingsley, from whom my father had expected the most instant appreciation, wrote that he could not . . . 'believe that God has written on the rocks one enormous and superfluous lie.' . . . a gloom, cold and dismal, descended upon our morning tea cups."

As Harold Morowitz points out in his article "Navels of Eden" in *Science 82* (March 1982), Philip Gosse was acquainted with Thomas Huxley and elected to the Royal Society for his work on animals called rotifers. He had met Charles Darwin, and over a period of many years exchanged friendly letters with Darwin about matters concerning plants and animals. "Not a word passes about evolution or creation," Morowitz writes, "or the enormous ideological gulf that separated the two great naturalists. The letters are quaint and polite and very British."

One of Edmund Gosse's best-known poems, "Ballad of Dead Cities," ends with the following stanza:

Henri Rousseau, *Eve*, 1905 (Art Resource, New York)

ENVOY

Prince, with a dolorous, ceaseless knell,
Above their wasted toil and crime
The waters of oblivion swell:
Where are the cities of old time?

Gosse could have written a poem about how the waters of oblivion dissolve even more rapidly such crank works as his father's effort to explain the fossil record.

I would have supposed that no creationist today could take *Omphalos* seriously. Not so! The *Des Moines Sunday Register* (March 22, 1987) published a letter from reader John Patterson arguing that the existence of a million-year-old supernova contradicted the notion that God created the entire universe about 4000 B.C. In its April issue, the newspaper ran the following response from a Donna Lowers:

> In regard to John Patterson's letter . . . on the supernova as a well-documented fact of science—of course it is! However, he cannot prove evolution except by circumstantial evidence, and creationists cannot prove creation except by God's word.
>
> To be a Christian requires one important element called faith. . . .
>
> Yes, I believe in creation by God in six days! I also believe in one day He created full-grown trees that contained rings that any scientist would declare had been there for years. He created pockets of oil deep in the earth that nature would take millions of years to process. He placed aquatic fossils far inland, and He created exploding stars for us to marvel about in the 20th century. . . .

Although few creationists today accept the thesis of *Omphalos,* a form of Gosse's argument is frequently invoked by young-Earthers to explain why the speed of light seems to prove the existence of galaxies so far from Earth that it has taken the light millions of years to reach us. God created the universe, they insist, with light from these distant galaxies *already on the way!* Gosse would have been delighted with this argument had he known about galaxies. Indeed, I myself like it better than the alternate conjecture that in the past light traveled millions of times faster than it does now.

As for the problem of navels, today's young-Earth creationists, who believe God fabricated Adam from the dust of the earth and Eve from Adam's rib, are strangely silent about the pair's navels. Silent, too, about other aspects of life that imply past histories. Would the trees in the Garden of Eden, for example, show rings if their trunks had been sliced? How would Jerry Falwell and other televangelists answer such questions?

Many liberal Christians, both Catholic and Protestant, now accept the evolution of the bodies of the first humans. However, as the present pope emphasized in his recent declaration that evolution is a legitimate theory, one must insist that God infused immortal souls into Adam and Eve—souls not possessed by their apelike ancestors. This is now the opinion of almost all leading Catholic thinkers. It forces the belief that the first humans, whether one pair or more than two, were reared and suckled by mothers who were soulless beasts. I once wrote a story about this titled "The Horrible Horns"—the horns are the horns of a dilemma—that you will find in my collection *The No-Sided Professor* (Prometheus Books, 1987).

Belly buttons are the topic of many old jokes, so let me end this column on a lighter note. It has been suggested that navels are most useful as a spot to put salt when lying on your back in bed and eating celery. And, an officer tells a civilian he's a naval surgeon. "Goodness me," the man replies, "how you surgeons specialize!"

CHAPTER 2

Phillip Johnson
on Intelligent Design

In November 1996 more than 160 scientists and scholars converged
on Biola University, in La Mirada, California, for the first annual conference of a movement called *intelligent design*. Its promoters are theists with
views ranging from conservative Christianity to a philosophical theism un-
connected with any religion.

Intelligent designers must not be confused with ignorant Christian fundamentalists who persist in believing that Earth and all its life were created
about ten thousand years ago, in six literal days, and that fossils are relics
of life destroyed by a worldwide flood. Many proponents of intelligent design (ID) have no quarrel with an ancient Earth. They accept the fact that
life evolved over millions of years from simple one-celled forms in Earth's

primeval seas. Their quarrel is only with the notion that evolution occurred without God's guidance.

On the other hand, many associated with the ID movement are unashamed "young-Earthers." Paul Nelson, who edits the ID newsletter *Origins and Designs,* is a fervent young-Earther, as is Nancey Pearcey, a featured speaker at the Biola conference. "Old-Earthers" are embarrassed by these fundamentalists in their midst, but do their best to downplay their influence.

Two developments in modern cosmology have played strong roles in the rise of the movement. The big bang suggests a moment of creation in which the entire history of the universe, including the eventual emergence of you and me, existed potentially in the properties of a small number of fundamental particles and their fields. The other driving force is the strong anthropic principle. It asserts that the universe could not have permitted life, or even the formation of suns and planets, unless some dozen basic constants of nature were extremely fine-tuned. In brief, IDers argue that modern cosmology implies a transcendental Designer. As physicist and pantheist Freeman Dyson memorably put it, "The universe in some sense must have known that we were coming."

IDers go much further than this. In a raft of impressive books, including the recent *Darwin's Black Box* (1996) by the Roman Catholic biochemist Michael Behe (his name rhymes with *tee-hee*), they contend that Darwinism has died. By "Darwinism" they mean the belief that evolution operates solely by random mutations and natural selection. In a narrow sense, of course, Darwinism long ago was modified by the discovery of mutations. The modern theory of evolution incorporates genetics and all other relevant findings of twentieth-century science. Darwin was a Lamarckian who accepted the now-abandoned notion of the inheritance of acquired traits.

In the last few years, many prominent political conservatives have defended ID. Irving Kristol, a firm believer in the God of Israel, has been attacking Darwinism for decades. His views are shared by his wife, Gertrude Himmelfarb, who in 1959 even wrote a biography of Darwin. Robert

Bork, in *Slouching Towards Gomorrah* (1996), cites Behe as having proved that "Darwinism cannot explain life as we know it. . . . Religion will no longer have to fight scientific atheism with unsupported faith. The presumption has shifted, and naturalistic atheism and secular humanism are on the defensive."

In its June 1996 issue, the conservative journal *Commentary* featured "The Deniable Darwin," a spirited defense of ID by David Berlinski, a mathematician who more recently published a popular introduction to calculus. "An act of intelligence is required to bring even a thimble into being," he wrote; "why should the artifacts of life be different?"

Pat Buchanan, a right-wing Catholic, denies evolution entirely. He has attacked it in his newspaper columns and, echoing William Jennings Bryan, said, "You may think you're descended from a monkey, but I don't." Of course, humans are not descended from monkeys, a common misconception.

The most influential book defending ID is Phillip E. Johnson's *Darwin on Trial* (InterVarsity Press, 1991, revised 1993). William Buckley had Johnson on his television show in 1989, and his book was reviewed with high praise in Buckley's *National Review* (April 19, 1991). The same magazine (April 22, 1996) allowed Johnson to do a hatchet job on Carl Sagan's *The Demon-Haunted World.* The equally conservative *New Criterion* (October 1995) let Johnson blast Darwinian Daniel Dennett's *Darwin's Dangerous Idea.*

Obviously I cannot comment here on all the many recent books by IDers, so let me focus on *Darwin on Trial.* Johnson is a mild-mannered, affable law professor at the University of California at Berkeley. He should not be confused with Philip (one *l*) Johnson, the eminent architect who designed Manhattan's AT&T building and televangelist Robert Schuller's Crystal Cathedral, in Garden Grove, California.

Although today's evolutionists all agree on the fact of evolution, they squabble over the mechanisms by which it operates. One major rift is between the gradualists, who follow Darwin in emphasizing slow change, and the "jump" theorists, notably Stephen Jay Gould, who stress long periods of stasis for many life forms, broken by periods of rapid change. By "rapid" they mean changes taking place by incremental mutations over tens of thousands of years—mere blips in geologic time.

Johnson is good in detailing these controversies. He takes them to bolster his view that there are dark mysteries about how life evolved—wide gaps that can be filled only by creative acts of God. He fully grants that random evolution occurs trivially within a species—the diversity of dogs, for example—but denies that new species can arise unless somehow directed from above. His fundamental claim, one stressed by all opponents of evolution from Darwin's day to now, is that structures as complicated as eyes and wings have no survival value unless they appear suddenly and fully formed. Intermediate stages, he falsely insists, are not in the fossil record and simply did not exist.[1]

It is strange that nowhere in his book does Johnson mention the British biologist St. George Mivart. Mivart spent a lifetime trying to persuade his church—he was a liberal Catholic—that its opposition to evolution was as monstrous a blunder as its earlier opposition to Galileo. Mivart argued in *The Genesis of Species* (1871), a book Darwin took seriously, that God's help is needed to explain transitions to new species, and especially for infusing an immortal soul into the first human bodies. All of Johnson's fundamental objections to Darwinism are in Mivart's old book. Mivart was excommunicated and denied a Christian burial. Ironically, his approach to evolution has now been officially endorsed by Pope John Paul II and is held by almost all Catholic theologians.

Mivart was the first major scientist to emphasize that eyes and wings are too complex to have evolved by slight modifications, and that such structures must appear suddenly because earlier incipient stages would have no survival value.

[1]There are thousands of "missing link" fossils, and every year more are found. Examples are the stages between reptiles and mammals, between reptiles and birds, between land mammals and whales, between horses and their progenitors, and between humans and their extinct apelike ancestors. The so-called fossil "gaps" are partly due to the rarity of conditions for fossilization and to the relatively rapid series of mutations emphasized by Gould and his associates.

From Mivart's day until now, creationists of all stripes have monotonously asked, "What use is half a wing?" In his popular book *The Blind Watchmaker*, Richard Dawkins answered as follows:

There are animals alive today that beautifully illustrate every stage in the continuum. There are frogs that glide with big webs between their toes, tree-snakes with flattened bodies that catch the air, lizards with flaps along their bodies; and several different kinds of mammals that glide with membranes stretched between their limbs, showing us the kind of way bats must have got their start. Contrary to the creationist literature, not only are animals with "half a wing" common, so are animals with a quarter of a wing, three quarters of a wing, and so on. The idea of a flying continuum becomes even more persuasive when we remember that very small animals tend to float gently in air, whatever their shape. The reason this is persuasive is that there is an infinitesimally graded continuum from small to large.

Similar arguments, detailed by Darwin himself, give equally plausible conjectures about how eyes could slowly evolve independently, in many different species, from light-sensitive spots on the skin. Although Johnson quotes Dawkins's scenarios for the gradual development of eyes and wings, he calls them speculative "fables" having no supporting evidence: "No one has ever confirmed by experiment that the gradual evolution of eyes and wings is possible." Dawkins would strongly disagree. He has an eloquent chapter on the multiple evolutions of the eye in *A River Out of Eden* (1995).

I found it also curious that Johnson never refers to the Dutch botanist Hugo de Vries, the man who coined the word *mutation*. De Vries argued persuasively, and for a brief time won many followers, that every new species appears "suddenly" as a result of a single macromutation in one generation.

When I finished Johnson's book, I was less interested in his moth-eaten objections to Darwinism than in what he would put in its place. On this all-important matter he is infuriatingly silent. There are four possibilities:

1. Johnson agrees that evolution operates by slight mutations followed by natural selection, but thinks God produced all the favorable mutations. I rule this out because Johnson constantly stresses the "abrupt" appearances of new species, with no transitional earlier forms.

2. Johnson thinks that new species, of which there are millions, emerged suddenly as a result of God-caused, massive mutations.

3. Johnson thinks that God intervened only in the creation of life and in producing massive mutations for broad classifications of life, such as plants, reptiles, mammals, fish, birds, and, of course, humans.

4. Johnson thinks that, at spots in the history of life, God created by fiat new life forms that had no ancestors. This is the view defended by many creationists who accept an ancient Earth but want to follow Genesis (assuming each "day" to be a long period of time) in having God perform thousands of miracles along the way.

In his books and writings, Johnson has steadfastly refused to explain how he thinks evolution crossed all those mysterious gaps in the fossil record. He sheds no light on what he thinks occurred when the gap between human and apelike creatures was bridged. Were there an Adam and Eve, or many Adams and Eves, created from the dust of the earth as Genesis tells? (In one of the Bible's two versions of creation, Eve was fashioned from Adam's rib, a miracle taken to be literally true by Jerry Falwell and other fundamentalists.) Or did God merely infuse souls into soulless animal bodies?

Please, Mr. Johnson, answer plainly some simple questions. Did the first mouse have a mother? Did the first humans have navels? If so, were they suckled and reared by beasts? Why did God put nipples on males?

It seems to me unfair for Johnson to lambaste evolution so fiercely without letting us in on what he thinks, or at least suspects, should replace it. It's like writing a book denying that Earth is round but never indicating what shape you think it is. Hoping to gain some answers to these questions, I exchanged a dozen letters with Johnson. He flatly refused to tell what variety of creationism he espoused. His reason? *Darwin on Trial* was intended only as an attack on godless Darwinism. He saw no need to reveal what should go in its place.

In a 1992 lecture recorded on videotape, replying to a question, Johnson admits that IDers have contradictory opinions on how God intervened. He expresses his hope that after Darwinism has been thoroughly discredited, as he is certain it soon will be, a "paradigm shift" will occur and scientists will be free to search for empirical evidence of just when and how God shoved evolution along. It's possible, I suppose, that Johnson has no strong opinion about this question.

I did discover that Johnson is an evangelical Presbyterian. But how far does he go in accepting New Testament miracles? I wrote to ask if he believed in the virgin birth, that Jesus raised Lazarus from a decaying corpse, or walked on water, or turned water into wine. Again, he refused to answer, though he did say he believed in Jesus' Resurrection, and that acceptance of other Biblical miracles offered no problem.

Johnson's second book, *Reason in the Balance* (InterVarsity Press, 1995), is mainly an attack on atheism, though it includes a chapter in which he again bashes Dawkins. As in his earlier book, he never lets us know whether he thinks new species appeared as a result of massive mutations directed by the Lord, or whether he thinks God created life forms that had no ancestors.

I was startled by a footnote (p. 257) in which Johnson says he greatly admires my religious novel, *The Flight of Peter Fromm,* even though it is an attack on Christianity. He cannot, however, comprehend how I can hold a "naturalistic worldview" and at the same time profess a belief in God.

Easy. Unlike Johnson, I am not shy about disclosing my basic convictions. I believe, by an emotional leap of faith, in a "wholly other" deity, utterly inscrutable to our little finite minds. I believe there are truths as far beyond our grasp as calculus is beyond the grasp of a cat. Because I also regard God as immanent in nature, we can say metaphorically that God both created and upholds the universe. I do not believe in what I call the "superstition of the finger"—the notion, to me close to blasphemy, that God finds it necessary at intervals to abrogate natural laws by injecting a finger into the universe to tinker with it. Newton not only was sure that God created the universe and all its laws in six days, he also believed it was necessary for God to periodically adjust the paths of planets to keep the solar system operating smoothly.

If Johnson does not share this Newtonian belief, why is he unable to grant that chance, combined with natural laws, is God's method of creation? I suspect it is because of his hidden agenda to defend conservative Presbyterianism. Regardless of Einstein's animadversions, chance is not a dirty word. It is absolutely and beautifully essential in quantum mechanics.

I sometimes fancy that quantum laws provided the only way, or perhaps the best way, God could bring about a monstrous universe capable of generating, after billions of years, intelligent life. The amazing thing is that an unconscious watchmaker, oblivious of any overhead plan, can actually work so well. Otherwise you would not be here to read these words.

One of Darwin's bitter crosses was the never-failing Anglican orthodoxy of his wife. In spite of her tearful entreaties, he abandoned his Christian beliefs early on, and eventually, after the death of his daughter Anne, lost faith in God entirely. However, in 1860, a year after the publication of his *Origin of Species,* Darwin defended intelligent design in a letter to Asa Gray:

> I see no necessity in the belief that the eye was expressly designed. On the other hand, I cannot anyhow be contented to view this wonderful universe, and especially the nature of man, and to conclude that everything is the result of brute force. I am inclined to look at everything as resulting from designed laws, with the details, whether good or bad, left to the working out of what we may call chance. Not that this notion *at all* satisfies me. I feel most deeply that the whole subject is too profound for the human intellect. A dog might as well speculate on the mind of Newton.

I couldn't say it better.

By far the harshest criticism of *Darwin on Trial* is a review by Stephen Jay Gould in the July 1992 *Scientific American.* Another excellent review, by anthropologist Eugenie C. Scott, appeared in *Creation/Evolution,* vol.

13 (1993), pp. 36–47. She concludes: *"Darwin on Trial* deserves to be read by scientists, not for its scientific value, which is negligible, but for its potential social and political impact."

Johnson's latest book is *Defeating Darwinism by Opening Minds* (InterVarsity Press, 1997). An ad in the publisher's catalog quotes Michael Behe: "Phillip Johnson is our age's clearest thinker on the issue of evolution and its impact on society."

Addendum

David Berlinski's spirited attack on evolution (*Commentary,* June 1996), like Philip Johnson's earlier book *Darwin on Trial,* contains one huge glaring omission. Nowhere does he tell us what brand of creationism he supports.

Like Johnson, Berlinski seems to think that the punctuated evolution of Stephen Gould and his friends somehow damages the Darwinian view that all life evolved by gradual small changes. The jumps in Gould's theory are, of course, jumps only relative to the extremely long periods during which certain species remained stable. Trilobites for example. Gould's jumps are tens of thousands of years, arising from accumulated, tiny mutations within small, isolated populations. Darwin's bulldog, Thomas Huxley, was well aware of such jumps, and they provided fuel for creationists from Darwin's day until now. Indeed, every argument against evolution put forth by Johnson and Berlinski goes back for more than a century. Today they are repeated over and over by Protestant fundamentalists who believe God created the entire universe in six days, about ten thousand years ago, just as it says in Genesis.

Commentary (September 1996) devoted twenty pages to letters attacking and defending Berlinski, including a letter of mine. I ended my letter by asking Berlinski, "Do you think the first humans had parents who were beasts, or no parents at all?" In his fifteen-page reply to the letters, Berlinski commented on my letter this way: "As for Mr. Gardner's last question: for many years I have been puzzling over whether the first humans had parents; sad to say, I still have no answer." I find this remark astonishing. If the first humans had no parents, then they must have been created on the spot by Jehovah. One wonders if Berlinski is open to the possibility that Eve was created from Adam's rib?

Consider a week-old baby. It is less "human" than a week-old gorilla. There is no moment along the continuum of a baby's growth at which it suddenly becomes a mature person. The evolution of *Homo sapiens* is on a similar spectrum. If the laws governing evolution were made and are upheld by God, what need is there for God to poke a finger into the process? The head of the Vatican's astronomical observatory, on a television show about Galileo, said it well: "There was no magic moment. The whole thing is magic."

Berlinki's opinions grew more mystifying when he attacked cosmological evolution in his article "Was There a Big Bang?" in the February 1998 issue of *Commentary*. (I have not seen his earlier article "The Soul of Man Under Physics," in the January 1996 *Commentary*.)

Because, Berlinski argues, there are grave doubts about the red shift as a measure of the receding velocity of galaxies, there is equal doubt that the universe is expanding, therefore no sound reasons to assume the universe originated in a big bang. Presumably Berlinski prefers a static universe, either always the way it is or created at some moment in the past.

The "black box" in the title of Behe's book is the living cell. He believes it to be far too complex to have evolved without divine aid. Behe's central exhibit is the rotating flagellum of certian bacteria. He insists there is no way to conceive of incipient forms that could explain how the flagellum could have evolved by slow natural selection. Like Johnson and Berlinski, we never learn just how he thinks God helped evolution along. "Don't worry, Mike," Johnson wrote to Behe. "Even if the [*New York*] *Times* bashes you in their review, a cultural earthquake will take place in the United States on August 4 when they publish it."

Of course no such quake occurred.

For a recent thoroughgoing attack on intelligent design, with special focusing on Johnson and Behe, I recommend Robert T. Pennock's *Tower of Babel: The Evidence Against the New Creationism* (1999). The unusual aspect of this book is that Pennock, unlike most defenders of Darwinian evolution as a blind watchmaker, is a theist in the Quaker tradition. He sees no need to assume that God boosted evolution along by performing little miracles along the way, since all laws operating on the evolution of life were created and are sustained by a wholly other deity.

It is odd, Pennock writes, that creationists who take the Bible so seriously should see God as analogous to humans by having him constantly tinkering with the universe in a way similar to how humans keep improving their cars, ships, and airplanes. He reminds them of Isaiah (55:8): "For my thoughts are not your thoughts, neither are your ways my ways, says the Lord."

In a column in the *Wall Street Journal* (August 16, 1999), Johnson referred to a "Chinese paleontologist" who "lectures around the world saying that recent fossil finds in his country are inconsistent with the Darwinian theory of evolution." As reported in *Skeptical Inquirer* (November/December 1999), physicist David Thomas wrote to Johnson to ask who this mysterious scientist was, and had he published any papers on the fossil findings? Johnson refused to name the man, adding that as yet he had published nothing in English. "My jaw dropped," said Thomas. "I'd expect a Deep Throat in politics—but not in science."

Behe's central argument, that atheistic evolution cannot explain the "irreducible complexity" of even a living cell, has been picked up by numerous defenders of ID. An example is engineer David Foster's book *Does God Exist?* (2000). See also his naive two-part article "Proving God Exists," in *The Saturday Evening Post* (November/December 1999 and January/Febraury 2000).

Behe's speculations also underlie *No Free Lunch: Why Specified Complexity Cannot Be Purchased Without Intelligence* (2002), by William A. Dembski. For criticism, see H. Allen Orr's review in the *Boston Review* (Summer 2002), and its references.

Part II

Astronomy

CHAPTER 3

Near-Earth Objects

Monsters of Doom?

Asteroid! Asteroid!
Speeding through the sky.
Will it strike or graze the earth?
Will everybody die?
　　　　　　　　　—Armand T. Ringer

In March 1998 astronomer Brian Marsden, at the Harvard-Smithsonian Astrophysical Observatory in Cambridge, Massachusetts, issued a spine-chilling announcement. Based on eighty-eight days of observing asteroid 1997 XF11, his computer calculated that this massive rock would come perilously close to Earth at 1:30 P.M. Eastern time, on Thursday, October 26, 2028. It could miss Earth by a mere 30,000 miles, only about one-eighth our distance from the moon. If the rock, almost a mile wide, struck the earth the devastation would be too awful to contemplate.

The next day, before fundamentalist cult leaders had time to integrate this possible cataclysm into their Second Coming prophecies, Marsden was humbly apologizing. Eleanor Helin and her associates at NASA's Jet Propulsion Laboratory located a seven-year-old photo of XF11 that permitted a more precise calculation of its path. The asteroid circles the sun every twenty-one months. On its 2028 crossing of Earth's orbit it will miss us by 600,000 miles, about two and a half times the moon's average distance from Earth.

"Near-earth objects" (NEOs) is today's term for massive objects that periodically cross Earth's orbit not far from our planet. They include asteroids, meteoroids that are mostly asteroid fragments resulting from collisions, and comets that come from regions far beyond Pluto. Disasters caused by NEOs striking Earth were common themes in early science fiction as well as in some modern disaster movies.

As usual, H. G. Wells pioneered the theme.[1] His novel *In the Days of the Comet* concerns the effects on Earth of a near miss by a giant comet. His short story "The Star" is a vivid account of devastation caused by a mammoth NEO. An asteroid (Wells calls it a "planet") from the outskirts of the solar system is shifted from its orbit. It collides with Neptune. The two coalesce to form a flaming "star" that almost demolishes Earth before it plunges into the sun.

Wells's story first ran in the 1887 Christmas issue of the London periodical *The Graphic*. I have framed in my study the magazine's full-page, full-color illustration showing Londoners staring upward at the star and shouting, "It is brighter!" A newsboy holds a paper with the headline "Total Destruction of the Earth" in huge scarlet letters.

Here is how Wells describes what happens as Earth and star swing around each other:

---

[1]This is not strictly true. There were a few earlier but undistinguished stories about Earth's encounter with NEOs; for example, Edgar Allan Poe's "The Conversation of Eiros and Charmian" (1839). Two former Earthlings, now disembodied spirits, recall Earth's destruction by a giant comet. It withdrew from Earth's atmosphere all its nitrogen. The remaining oxygen caused Earth to explode in flames. Everett Bleiler's *Science Fiction: The Early Years* (1990) lists "The Comet," a story by S. Austin, Jr., also published in 1839, in which Earth is destroyed by a comet.

And then the clouds gathered, blotting out the vision of the sky, the thunder and lightning wove a garment round the world; all over the earth was such a downpour of rain as men had never before seen, and where the volcanoes flared red against the cloud canopy there descended torrents of mud. Everywhere the waters were pouring off the land, leaving mud-silted ruins, and the earth littered like a storm-worn beach with all that had floated, and the dead bodies of the men and brutes, its children. For days the water streamed off the land, sweeping away soil and trees and houses in the way, and piling huge dykes and scooping out Titanic gullies over the countryside. Those were the days of darkness that followed the star and the heat. All through them, and for many weeks and months, the earthquakes continued.

I still recall as a boy reading in Hugo Gernsback's *Science and Invention,* then my favorite magazine, the six installments of a preposterous novel that ran in issues July through December 1923. The author was Ray Cummings, and his novel was titled *Around the Universe: An Astronomical Comedy.* It was about a spaceship that carried Tubby, his girlfriend, and an astrophysicist called Sir Isaac. After exploring the universe, they learn that evil Martians are planning to invade the earth. To prevent this, Sir Isaac steers his spaceship in circles around a tiny asteroid, causing it to shift its orbit slightly. This results in a series of collisions with larger asteroids, all precisely calculated by Sir Isaac, that finally create a giant fireball which collides with Mars and obliterates its inhabitants.

This is not as crazy as it seems. The orbits of asteroids are chaotic. A minute change in one orbit could start a "butterfly effect" of the sort triggered by Sir Isaac. Even our solar system is unstable. A giant NEO coming close to a small planet or striking it could set off a chain reaction that could conceivably kick a planet out of the solar system. (See "Crack in the Clockwork," by Adam Frank, in *Astronomy,* May 1998, pp. 54–59.) Newton was well aware of this instability. He believed it was necessary for God to intervene at times to readjust planetary orbits to keep the system running smoothly.

Later science fiction novels about NEOs that wreck Earth are far too numerous to list. On the movie screen New York City has twice been destroyed by NEOs. It was demolished in a dreary 1979 film, *Meteor,* which wasted the talents of Sean Connery, Natalie Wood, Henry Fonda, and Trevor Howard. In an earlier and even more absurd film, *When Worlds Collide* (1951), a wandering star called Ballus flattens New York City with a gigantic tidal wave. The recent XF11 scare has been great publicity for two new disaster films about NEO impacts released this year just after the time I am now writing: Disney's *Armageddon,* starring Bruce Willis, and Paramount's *Deep Impact,* featuring Robert Duvall. It's a good bet that the visual effects of both films will be superior to their scientific accuracy.

When Worlds Collide was based on a popular 1933 novel by Philip Wylie and Edwin Balmer. In the book the doomsday NEO is called Bronson Alpha—a huge planet around which a small earthlike planet (Beta) whirls. The pair have broken away from a far-distant sun system. Beta once supported a culture of sentient beings, and the cold of outer space has, of course, frozen everything on its surface.

Alpha's close brush with Earth destroys its cities with violent tidal waves, earthquakes, and volcanic eruptions. As the authors put it: "The Earth burst open like a ripe grape." There are survivors. The planets loop around the sun, then return to Earth. After destroying the moon, Alpha collides with Earth, reducing it to fragments. Alpha then leaves the solar system on a hyperbolic path.

Beta remains to replace Earth as a permanent member of the sun's system. Just before Earth vanishes, a group of several hundred brave men, women, and children manage to escape in two atomic-powered spaceships. They land on Beta, where they find a warmed-up planet with an emerald sky, surviving plants, and a breathable atmosphere almost identical to Earth's. They will colonize the planet and preserve humanity. Lovers Tony and Eve provide the book's romantic interest.

When the novel began serialization in *Blue Book* magazine in 1932 it was an instant sensation among SF buffs. One critic called it "an astronomical fantasy of the first magnitude." The hardcover became a best-seller. In my opinion it is low-level pulp fiction with little hard science to redeem it.

The sequel, *After Worlds Collide,* which began in *Blue Book* in 1933 and was published as a book the following year, is even worse. The colonizers find plastic-covered cities on the planet, metal roads, curious airships, and a vast electric power station. There is no trace of inhabitants. Paintings show the denizens to be humanoid. Other earthlings have also escaped in spaceships to Beta. A war breaks out between the Americans and a group of Asiatic Communists. The Americans win. A great mystery remains. What happened to Beta's humanoids? A second sequel was contemplated to answer the question, but Balmer and Wylie failed to agree on the plot and it was never written.

Later science-fiction authors have been less concerned with NEO collisions than with adventurers who mine the asteroids for iron, nickel, and more valuable minerals. Examples are Clifford Simak's story "The Asteroid of Gold," and Malcolm Jameson's "Prospectors of Space." Other tales involve asteroids that are disguised as spaceships, or being used as stopover spots for explorers on their way to more distant reaches of the solar system.

Asteroid comes from a Greek word meaning "starlike." They were named that because early telescopes could see them only as points of light. Two large asteroids have since been photographed up close by space probes. They resemble misshapen potatoes, their surfaces pockmarked with craters like the surface of our moon.

The asteroid belt, between the orbits of Mars and Jupiter, contains tens of thousands of asteroids with diameters of a mile or more. The larger ones are spherical, but smaller ones, their cohesion greater than their gravity, are extremely irregular. There is no lower limit to asteroid size because they grade down to tiny rocks and particles of dust. None is big enough to hold an atmosphere.

Ceres, the largest asteroid and the first to be discovered (in 1801) is more than 600 miles across. A year later Pallas, about 370 miles wide, was found. Juno, about 140 miles across, and Vesta, with a diameter of possibly 330 miles, were found in 1804 and 1807 respectively. Because of its surface's high reflectivity, Vesta can at times be seen by unaided eyes. The four asteroids combined account for more than half the total mass of all the asteroids. Altogether they would make a planet smaller than our moon.

Ida, the second asteroid to be photographed close up (Gaspara was the first), is about thirty-six miles long. To the amazement of astronomers, the photo revealed a tiny moon orbiting Ida, possibly a piece chipped off the asteroid. Several other asteroids are believed to have a moon. In 1999 a moonlet was photographed orbiting an asteroid called Eugenia. A probe called "Near Earth Asteroid Rendezvous" (NEAR) will soon be snapping close-up pictures of Eros, the largest of the NEOs, which crosses Earth's orbit every forty-four years.

Asteroids were once thought to pose a grave hazard to space flight, but it turned out that distances between those large enough to damage a spaceship are so vast—typically millions of miles—that the danger of such accidents is near zero. Fatal collisions of spaceships with asteroids were common in early space operas before the asteroid belt was found to be much cleaner than previously suspected.

For a while the asteroids were named after Greek gods (at first, only females), but when such names ran out, they began to be named for cities, states, nations, persons (real and fictional), and even pets. Whoever found a new asteroid was usually allowed to name it. Today, most asteroids are designated by the year of discovery followed by letters and maybe a number. I am pleased to report that James Randi and I have asteroids named for us. CSICOP and its founder Paul Kurtz both had asteroids named for them on CSICOP's twentieth anniversary (*Skeptical Inquirer,* September/October 1996, p.8).

What produced the asteroids? In Conan Doyle's novel *The Valley of Fear* we learn that Sherlock Holmes's bitter enemy, Professor Moriarty, wrote a treatise titled "The Dynamics of an Asteroid." Isaac Asimov once conjectured that this obscure paper argued that asteroids are remnants of a small planet whose inhabitants discovered nuclear energy and blew their world to smithereens. This notion, once a favorite of science-fiction writers, has been abandoned on the grounds that not even a nuclear explosion would be great enough to form the asteroid belt. The prevailing view is that the rocks are material that failed to coagulate into a planet, perhaps because of the strong gravitational influence of nearby Jupiter.

There is no doubt that eventually Earth will be struck by a massive NEO, because such events have often occurred in the past. The most recent was the 1908 crash of a large NEO in the Tunguska River Valley of central Siberia. It flattened trees for many miles around and killed a herd of reindeer. Almost two hundred impact craters have been identified that testify to similar impacts, and there surely are thousands of craters that vanished long ago from erosion. It is widely believed that the impact of a giant NEO caused a mass extinction of life, including the dinosaurs, 65 million years ago at the end of the Cretaceous era.

Almost all asteroids are confined to the asteroid belt, but many wander far beyond the orbit of Jupiter, and others plunge inward past the orbit of Venus. One called Icarus swings inside Mercury's orbit, and Charon floats beyond Saturn. The two moons of Mars may be captured asteroids. It is estimated that more than a thousand asteroids at least a mile wide are NEOs. Perhaps a dozen are three or more miles wide. They pose a monstrous threat to humanity if they come close to Earth or hit it.

In 1937 Hermes, half a mile wide, missed us by about twice the distance from Earth to the moon. In 1989 an asteroid called Asclepius, also about half a mile across, came even closer. In 1991 a small asteroid about thirty feet wide missed the earth by less than half the distance to the moon. The latest near miss was in 1996 when JA1, a third of a mile wide, set a record for large asteroids by missing us by a mere 280,000 miles, only forty thousand miles more than the Earth-moon distance.

If some time in the future an asteroid is determined to be on a near collision with Earth, what can be done to prevent disaster? One suggestion, not overlooked in science fiction, is to attach a nuclear bomb to the rock that will blow it into a harmless orbit. (Early science fiction used cannonballs to deflect comets.) The danger of this is that it could produce fragments that would hit the earth, causing even more damage than the intact rock.

This is exactly what happens in *Asteroid,* a four-hour 1977 television film which NBC aired again in March, 1998. A comet alters the orbits of a group of asteroids and sends them hurtling toward Earth. The largest, Eros, is exploded with laser beams, but the thousands of resulting fragments are large boulders that rain down on Earth to cause unspeakable devastation. The film's visual effects are riveting, especially the leveling of Dallas skyscrapers, but the search for survivors and the rescue scenes seem interminable. Better techniques for diverting an asteroid may be landing a rocket engine on the rock to nudge it into a harmless trajectory, or attaching a large solar sail to let the sun's radiation do the nudging.

Suppose, however, there is not enough time for measures to be taken to prevent a collision, and Earth is shattered by a giant NEO that will hurl us all into oblivion. What are the philosophical implications of such an event? This obviously is not a problem for atheists, agnostics, or pantheists because they are resigned to the fact that Nature does not care a rap about preserving a species.

What about theists? I'm inclined to think that even to them a sudden extinction of humanity would be acceptable. The Biblical Jehovah, remember, is said to have drowned every man, woman, baby, and their pets, except for Noah and his family.

If God can allow an earthquake to kill thousands, or the Black Death to wipe out half of Europe, surely she would have no scruples about allowing an asteroid to bring human history to a flaming end.

References

"Is the Sky Falling?" by planetary astronomer David Morrison, in *Skeptical Inquirer,* May/June 1997. Morrison reviews ten recent trade books on the topic, including three he considers worthless.

"Collisions with Comets and Asteroids," by Tom Gehrels, in *Scientific American,* March 1996.

On the miscalculation of XF11, see "Whew!" by Leon Jaroff, in *Time;* "Never Mind!" by Adam Rogers and Sharon Begley, in *Newsweek;* and "Okay, So They Were a Little Off," by Charles Petit, in *U.S. News and World Report.* All three articles were in March 23, 1998, issues.

Hazards Due to Comets and Asteroids, a collection of papers edited by Gehrels, who runs an asteroid search program at the University of Arizona. This standard reference was published by the University of Arizona Press in 1994.

"Space Opus: Philip Wylie," Chapter 17 in *Explorers of the Infinite: Shapers of Science Fiction* (1963), by Sam Moskowitz.

Addendum

I took notes after seeing *Deep Impact* and *Armageddon.*

Impact features an enormous comet headed toward Earth. A spaceship called *Messiah,* armed with nuclear bombs, is sent to blow up the comet. Unfortunately, all the bombs do is blast the comet into two parts, both still hurtling earthward. Efforts to divert the parts fail. The smaller one lands in the Atlantic, creating a huge tidal wave that destroys New York City, Washington, D.C., and Philadelphia.

The *Messiah*'s crew, captained by aging astronaut Robert Duvall, remains in space. The only way to save Earth from total destruction by the larger of the two chunks is to plunge the *Messiah* into the chunk and explode the ship's remaining nuclear bombs.

On their way to fulfil this suicide mission, the film's most implausible scene occurs. An astronaut on board has been blinded by flames that erupted on the comet while the bombs were being planted. Duvall has brought along a copy of *Moby-Dick.* He starts reading it to the blind man: "Call me Ishmael." Evidently the script writers intended the comet, like Melville's white whale, to be a symbol of evil. The suicide mission is successful. The big chunk is blasted into fragments that shower harmlessly down on the rescued Earth.

Meanwhile the tidal wave produced by the smaller chunk has killed millions around the world. The movie ends with the U.S. president, a handsome black man, making a televised speech in which he thanks God for humanity's survival and vows that the demolished cities will be rebuilt.

I spare readers from detailing the film's two subplots. One involves a young astronomer and his girlfriend. The other concerns a TV anchorwoman who is alienated from her father. They are reconciled in a touching scene in which they stand on a seashore, embracing one another, as they watch the high tidal wave approach them.

Armageddon, consisting mainly of spectacular visual effects accompanied by ear-blasting sound, was noisier than a rock concert; not just the explosions, collisions, and impacts, mind you, but astronauts incessantly shouting at one another. Experts decide that the only way to save Earth from an asteroid the size of Texas is to plant a nuclear bomb inside the monster. To accomplish this they call on Bruce Willis, the world's greatest oil well driller.

Harry, as Willis is called, has caught his only child, Grace, making love with one of his roughnecks. He is so furious that he even tries to shoot the man. However, the poor chap is essential to take along on the mission. The crew lands on the asteroid. Large asteroids are round and smooth, but this one has exotic scenery with large spikes that stick up from the surface. The crew drill a deep hole into which they lower the bomb. Alas, the detonation device, which they intend to operate as soon as they are safely back on their spaceship, malfunctions. Someone has to stay behind to explode the bomb by hand.

Crew members draw straws. Grace's lover gets the short one. He is eager to stay behind to detonate the bomb, but Harry, who has grown to admire the fellow, knocks him flat. Brave Harry then remains on the asteroid to blow it in half and himself to smithereens. The two halves of the asteroid will then pass harmlessly on opposite sides of the earth.

The picture ends with Harry on television, just before his sacrificial death, telling his daughter how much he loves her and how great her boyfriend is. Armageddon is averted, thanks to modern technology and the altruism of Harry, master oil driller.

Only the film's perpetual noise kept me from snoozing. To make everything politically correct, Harry's crew includes an African American, and a woman who seems to do nothing until near the end when she springs into action to get the spaceship started back home.

Everett Bleiler's *Science-Fiction: The Gernsback Years* (1998) turns up two NEO tales worth mentioning.

"The Falling Planetoid," by Isaac Nathanson (*Science Wonder Stories,* April 1930), tells how an asteroid headed for Earth is turned into a harmless second moon by a series of explosive shots.

"The World of a Hundred Men," by Walter Kateley (*Science Wonder Stories,* March 1930), is about an excavation of the meteor crater in Arizona. An unearthed object turns out to be a small planetoid on which humanoid life flourished. Knowing their unavoidable collision with Earth, the hundred ill-fated inhabitants have built a durable museum to preserve their history.

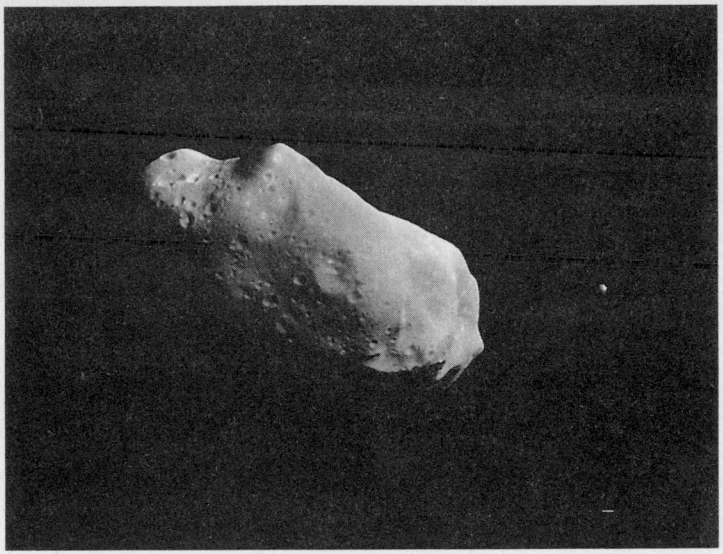

Asteroid 243 and its tiny moon. (NASA/JPL)

Ron Miller called my attention to *The Moon Maker* (1916) by Arthur Train and the famous American physicist Robert Wood. A huge asteroid called Medusa is diverted from collision with Earth by rays that produce a nuclear reaction on its surface. Medusa then becomes Earth's second moon.

Doomsday Rock, a 1997 movie that I saw on a TV rerun, was a shameless pandering to the public's obsession with the occult. All astronomers agree that an asteroid called Nemesis will miss the Earth by a wide margin, but an anthropologist turned astronomy professor knows better. He has discovered a cave in Australia where thousands of years ago an Aborigine psychic painted pictures that accurately predict major events. These events end with Earth's destruction by a "demon rock" from the skies. Sure enough, a comet mysteriously splits, part of it hits Nemesis sending it straight toward Earth. A few seconds before impact, the professor, previously thought mad by his colleagues, saves the Earth by destroying the asteroid with a nuclear armed rocket.

CHAPTER 4

The Star of Bethlehem

Now when Jesus was born in Bethlehem of Judea in the days of
Herod the king, behold there came wise men from the east to
Jerusalem, Saying, Where is he that is born King of the Jews? for
we have seen his star in the east, and are come to worship him.
. . . and lo, the star, which they saw in the east, went before
them, till it came and stood over where the young child was.
When they saw the star, they rejoiced with exceeding great joy.
—*Matthew 2: 1–2, 9–10*

Whenever Christmas approaches, Protestant and
Catholic churches celebrate the birth of Jesus, with many references in ser-
mons and Sunday schools to the Star of Bethlehem. The nation's some hundred planetariums devote Christmas programs to possible natural causes
of the Star. According to the Book of Matthew, the only gospel to give an
account of the Star, the wise men from the east (their number is not given,
but tradition makes it three) were guided westward by the Star to the stable where the newborn Jesus lay in a manger.[1] There was no room at the

[1]Matthew's account of the visiting magi is retold in greater detail in the apocryphal Book
of James, a Greek manuscript of the second century. Legend has it that it was written by

inn for his parents (perhaps I should say "parent" because the gospels make clear that Joseph was not the baby's father).

Saint Augustine and other early Catholic theologians took for granted that the Star was one of God's miracles, placed in the heavens to lead the wise men to Bethlehem. When Copernicus, Kepler, and Galileo ushered in the rise of empirical science, it became fashionable for Christian scholars to seek natural causes for events which the Bible clearly describes as supernatural.

One of the most popular and longest-lasting of natural explanations of the Star was put forth by Kepler. He suggested in a 1606 tract that the Star was actually a conjunction of Jupiter and Saturn that occurred in 7 B.C. in the constellation of Pisces, the Fish. He was not the first to suggest this; the conjecture can be found in English church annals as far back as 1285, but Kepler was the first to argue the possibility at length. The constellation's name was a happy coincidence because a fish had long been, as it still is today, a symbol of the Christian church and its believers.

Scholars now agree that Jesus was born sometime between 4 and 8 B.C. Matthew dates the birth as in the "days of Herod." Herod is known to have died early in 4 B.C. so Jesus must have been born before then. The exact year is, of course, unknown, though it could well have been at the time of the 7 B.C. Jupiter-Saturn conjunction.

Kepler later had doubts about his conjecture. As astronomer Roy K. Marshall points out in his booklet *The Star of Bethlehem* (published in 1949 by the Morehead Planetarium, at the University of North Carolina, Chapel Hill), Jupiter and Saturn, throughout the period of their proximity, were never closer together than two diameters of the moon as it appears in the sky. In 1846, British astronomer Charles Pritchard did some careful research on the event. Because of the erratic looping paths of the two planets, as seen from Earth, there were three separate close encounters. Astronomers call it a "triple conjunction."

---

a half brother of Jesus. According to Origen, he was one of Joseph's sons by a former marriage. Chapter 15, verse 7, describes the Star as so huge and bright that it rendered all the other stars invisible.

The two giant planets were closest on May 29, October 1, and December 5. "Even with . . . the strange postulate of someone with weak eyes," Pritchard wrote, "the planets could not have appeared as one star." Marshall adds: "Only an abysmally weak pair of eyes could have ever merged them."

There are other objections to Kepler's guess. A much closer meeting of the same two planets occurred in 66 B.C. As Arthur C. Clarke says in his entertaining essay "The Star of Bethlehem" (Chapter 4 in his collection of essays *Report on Planet Three,* 1972), this event "should have brought a delegation of wise men to Bethlehem sixty years too soon!"

Each of the three conjunctions of 7 B.C. lasted only a few days, whereas Matthew has the Star guiding the wise men throughout a journey that must have taken at least several weeks. Finally, the two planets would rise and set like ordinary stars, planets, and the sun and moon, but Matthew describes the Star as lingering in the sky as it glided slowly toward Bethlehem. Kepler eventually decided the Star was created by God between Jupiter and Saturn when they were close together.

Kepler's original conjecture became popular among nineteenth-century Christians, especially in Germany, where the so-called "higher criticism" of the Bible favored natural causes for Biblical miracles. The 7 B.C. theory was also defended in endless popular biographies of Jesus published in Christian countries. In England the Anglican cleric Frederic W. Farrar, in his *Life of Christ* (1874), devotes several pages to a scholarly discussion of the 7 B.C. conjunction. Samuel J. Andrews, in *The Life of Our Lord upon the Earth* (1891), an American work, also takes Kepler's theory seriously.

In recent years the 7 B.C. conjecture has been revived in the lengthy life of Jesus section that makes up the final third of the massive *Urantia Book* (1955). This bible of the Urantia movement purports to have been written entirely by supermortals who channeled the text through members of the movement to give to Urantia, the cult's name for Earth, a new revelation destined to supersede Christianity. On page 1352 of the *Urantia Book* we learn that the Jupiter-Saturn encounter of May 29, 7 B.C., gave the appearance of a single star, which we know it didn't, and this accounts for what the supermortals call the "beautiful legend" that grew up about the

"Star." The supermortals, or "unseen friends" as Urantians like to call them, reveal that Jesus was born at noon, August 21, 7 B.C. It is a date celebrated annually by Urantians. (For more on the bizarre Urantia movement see my book *Urantia: The Great Cult Mystery*, now in its third printing by Prometheus Books.)

Other planetary conjunctions in later years have been considered as possible explanations of the Star. For example, a spectacular merging of Jupiter and Venus took place on June 17, 2 B.C. The disks of the two planets actually overlapped! This candidate for the Star is defended by James De Young and James Hilton in "Star of Bethlehem" (*Sky and Telescope*, April 1973), and again by Roger Sinnott in "Computing the Star of Bethlehem" (*Sky and Telescope*, December 1986). Jupiter and Venus were last that close in 1818, and won't be again until 2065.

Still another contender for the Star is a supernoval explosion that occurred in the spring of 5 B.C. in the constellation of Capricorn. You'll find this argued by British astronomer David H. Clark and two associates in the *Quarterly Journal of the Royal Astronomical Society* (December 1977). Other speculations, too absurd to consider, have hung the Star on Venus, comets, exploding meteors, and even ball lightning.

Immanuel Velikovsky, an Orthodox Jew, struggled to invent natural causes for Old Testament miracles. He was not, of course, interested in doing the same thing for the New Testament miracles. He even proposed a natural explanation of how Joshua made the sun and moon stop moving: It was really Earth that ceased rotating. This was caused by a mammoth comet that erupted from Jupiter and passed close to Earth before it settled down to become Venus! Some of today's far-out New Agers who believe in the reality of PK (psychokinesis) regard Jesus as a great psychic who used natural psi powers to walk on water, multiply loaves and fish, turn water into wine, and perform other stupendous feats of magic.

Ellen Gould White, prophetess and one of the founders of Seventh-day Adventism, had a much simpler, and more sensible, approach to the Bible's great miracles. She took them to be miracles. In *The Desire of Ages*, her book on the life of Jesus, she explains the Star as follows:

The wise men had seen a mysterious light in the heavens upon that night when the glory of God flooded the hills of Bethlehem. As the light faded, a luminous star appeared, and lingered in the sky. It was not a fixed star nor a planet. . . . That star was a distant company of shining angels. . . .

The association of the Star with angels goes back to the early church fathers. Longfellow, in the third section of his miracle play "The Nativity" (it is part of his book *Christus: A Mystery*), toys with the notion that the Star was held in the sky by angels. There were seven: angels of the sun, moon, Mercury, Venus, Mars, Jupiter, and Saturn. Here is Longfellow's opening stanza:

> *The Angels of the Planets Seven,*
> *Across the shining fields of heaven*
> *The natal star we bring!*
> *Dropping our sevenfold virtues down*
> *As priceless jewels in the crown*
> *Of Christ, our new-born King.*

What is my opinion about all this? I find it hard to comprehend why conservative and fundamentalist Christians, who believe the Bible's miracles to be actual events, would even try to find natural explanations for what the Bible clearly describes as divine supernatural phenomena. The Jehovah of the Scriptures has awesome powers to suspend natural laws and do whatever he wants. Why trouble to look for natural causes of the great downpour by which God drowned every man, woman, and child on Earth, as well as their pets, except for one undistinguished family and the few animals they took on their Ark? I once asked a Seventh-day Adventist why God would be so cruel as to murder all the innocent little babies. He replied that God foresaw how wicked they would become if allowed to grow up!

In my not-so-humble opinion, the story of the Star is pure myth, similar to many ancient legends about the miraculous appearance of a star to herald a great event, such as the birth of Caesar, Pythagoras, Krishna (the Hindu savior), and other famous persons and deities. Aeneas is said to have been guided by a star as he traveled westward from Troy to the spot where he founded Rome. (I was unable to find a reference to this in Virgil's *Aeneid,* and would be grateful to any reader who can locate the reference for me.) The legend about the Star of Bethlehem is believed by many scholars to have arisen to fulfill a prophecy in Numbers 24:17, "I shall see him [God], but not now. I shall behold him, but not nigh: there shall come a star out of Jacob, and a Sceptre shall rise out of Israel."

Although I do not think the Star of Bethlehem ever existed, or was an illusion caused by a natural astronomical event, I find Mrs. White's statement more to be admired than the futile efforts of liberal Christians to banish from the Bible all references to God's miraculous powers. I find this almost as degrading of the Bible as the efforts of ultrafeminist Christian leaders to expunge from Scripture every sentence in which God is called "Father" (or given any other masculine term), and Jesus is called "Son." The practice strikes me as even more ridiculous than trying to change "nigger," in such classic novels as *Huckleberry Finn* and Joseph Conrad's *Nigger of the Narcissus,* to "African American."

Let the Bible be the Bible! It's not about science. It's not accurate history. It is a grab bag of religious fantasies written by many authors. Some of its myths, like the Star of Bethlehem, are very beautiful. Others are dull and ugly. Some express lofty ideals, such as the parables of Jesus. Others are morally disgusting. I think of the tragic legend about the rash vow of Jephtha that prompted him to sacrifice his daughter. (Why does Saint Paul speak of Jephtha as a man of great faith?) Or the account of how an angry Jehovah slew Moses' two nephews with lightning bolts merely because they failed to mix the incense properly for a sacrifice. God didn't like the way the smoke smelled! The Old Testament's God is as skillful as Zeus at using lightning as a weapon of punishment.

The King James Bible is itself a near-miracle, its poetic style far more beautiful and moving than any modern translation in English or any other language. It is also an improvement over the frequently crude writing by the old Hebrew and Greek authors. The King James Bible is a literary masterpiece best left unaltered. It is a classic to put on a shelf alongside the great fantasies of Homer, Virgil, Dante, Milton, and yes, even the Koran.

Addendum

In 1999, soon after this chapter was written, two books about the Star were published by university presses: *The Star of Bethlehem* by Mark Kidger (Princeton University Press), and *The Star of Bethlehem* by Michael Molnar (Rutgers University Press).

Kidger, a British astronomer, thinks the Star was a nova that Chinese astronomers recorded as blazing in the sky for seventy days in 5 B.C. It followed a series of conjunctions that the wise men took as astrological signs that the Messiah had been born.

Molnar, an astronomer at Rutgers, argues that the Star was a myth based on an astrological event in which Jupiter was occulted by the moon in the constellation of Aires on April 17, 6 B.C. This, he believes, was the date of Jesus' birth. Matthew incorrectly described this astrological event as a star that moved through the sky.

If Matthew was so wrong about a star, how can we be sure he was right about a journey of wise men from the east? Molnar's conjecture strikes me as just as irrelevant as the other guesses about an event in the sky that could be taken as a star. Surely the simplest explanation of Matthew's account is that both the Star and the Magi belong among the many gospel legends that have no factual basis.

Dennis J. Cuniff, David Barclay, and Don G. Evans wrote to answer my question about where to find the Star in Virgil's *Aeneid.* The passage begins in line 694 of Book 2. I erred in thinking the Star guided Aeneas to the site of Rome. It was simply an omen in the heavens, produced by Jupiter, to let Aeneas know that he favored Aeneas's plan to found a new city in Italy. Accompanied by a thunderbolt, the Star was a bright meteor that streaked across the sky, leaving a trail of light and the odor of sulfur.

However, I was not so wrong after all when I referred to a legend that a star guided Aeneas in his search for a place to found Rome. William C. Waterhouse, a Penn State mathematician, wrote to tell me about a passage written by Marius Terentius Varro, a Roman scholar. The passage is quoted in a commentary on the *Aeneid* written in the fourth century A.D. by a man called Servius:

> Varro says that this Morning Star, which is called the Star of Venus, was always seen by Aeneas until he came to the Laurentian land; and once he got there, it ceased to be visible, from which he recognized that he had reached his destination.

Apparently this refers not to a temporary star but to the planet Venus which, according to Varro, seemed to lead Aeneas to his destination before it became invisible in the sky.

Part III

Physics

CHAPTER 5

The Great
Egg-Balancing Mystery

There is not the slightest doubt that one's mind can exert strong, totally subconscious influences on tasks that involve the hands. It is the secret of the Ouija board. It is the secret behind the sudden turning of dowsing rods whenever the dowser crosses a certain spot on the ground. Novelty stores used to sell what they called a "sex indicator." It consisted of nothing more than a small weight attached to the end of a string. You can make one in a jiffy. Hold the string's free end, allowing the weight to hang. When the weight is held above a man's hand, it will swing back and forth in a straight line. Held over a woman's hand, it will swing in an elliptical orbit. This works, of course, only if the person holding the string knows what to expect. Subconscious hand movements cause the device to fulfill expectations.

A recent scandal based on the Ouija-board effect is the claim that some autistic children while aided by a "facilitator" will type long documents far beyond the children's capacities to communicate by speaking. It has been shown by ingenious tests that a facilitator subconsciously guides the autistic child's hands as the child hits the keys. There have even been cases when autistic children in the hands of neurotic facilitators have typed fake condemnations of horrible sexual abuse by their loving parents!

For more than a century magicians have located hidden objects by what in the trade is called "muscle reading." A person who knows where the object is concealed grasps the magician's wrist. Subconscious pressures by the person's hand guide the magician to the correct spot. (Some magicians, I should add, unwilling to take chances with an uncooperative spectator, will have a "stooge" in the audience send electronic signals by a reed switch in a shoe. A tiny receiving device on the magician's body produces pulses that tell him or her which way to go.)

One of the funniest examples of mind control over the body is the annual ritual in China of balancing fresh chicken eggs on their broad end on the first day of spring. The notion that the position of the sun or planets on a certain day can influence gravitational forces acting on the egg is so preposterous that physicists laugh at the theory. Yet intelligent people, unknowledgeable about science and inclined toward paranormal beliefs, actually think that at certain times of the year a fresh egg is more easily balanced than at any other time!

This egg-balancing ritual seems to trace back to ancient China. Tradition has it that on Li Chun, China's first day of spring (the name means "spring begins"), eggs will balance on a smooth surface with greater ease than on other days. Old Chinese books of uncertain date, such as *Secret Kaleidoscope* and *Know What Heaven Knows,* are sources of this legend.

The legend reached the United States in 1945 when an article by Annalee Jacoby, describing the Chinese ritual, appeared in *Life*'s March 19 issue. Like our Thanksgiving, Li Chun is a variable date. It usually falls on February 4 or 5. In 1945 it was February 4, the twenty-second day of the twelfth Chinese lunar month. Some years have no Li Chun. These are called "blind" lunar years because they fail to "see" the first day of spring. Other lunar years can have two adjacent Li Chuns.

According to the *Life* article, in 1945 most of the population in Chungking turned out on Li Chun to balance eggs. All over the city one could see fresh eggs, shells unbroken, balancing on pavement, tables, and other surfaces. Correspondents for the United Press wired back stories about the mania. Albert Einstein was reported to have said he doubted that the date had any influence on egg balancing. Chungking was divided between believers and skeptics. Someone proposed balancing a large number of eggs to spell "Einstein is nuts," but nothing came of it.

For reasons that reflect popular ignorance of science, combined with a love of miracles, the notion that fresh eggs balance more easily on the first day of spring caught fire in the United States. However, the first day of spring here is the day of the vernal equinox, when the sun crosses the equator and day and night are of equal length. This occurs about March 21, more than a month after China's first spring day. But this discrepancy did not trouble American believers.

Life's article touched off a small epidemic of egg balancing in the United States, not on Li Chun, but on the vernal equinox. The mania crested nearly forty years later, in Manhattan in 1983. According to a three-page report in *The New Yorker* (April 4, 1983), a believer named Donna Henes organized her sixth annual egg-balancing ceremony in the Ralph J. Bunche Park at First Avenue and 42nd Street, across from the United Nations building. On March 20 the sun crossed the equator at precisely twenty-one minutes before midnight. At that instant, Henes believed, eggs would balance easily on their wide end.

Henes was then a thirty-seven-year-old artist strongly committed to working for world peace. Her egg-balancing ritual was intended to promote international harmony. The event was heralded by setting off fifty-two highway emergency flares, one for each week of the year. While the flares burned, Henes distributed from a laundry basket 360 fresh eggs donated by the Jersey Coast Egg Producers. Why 360? Because, Henes explained, there are 360 degrees in the earth's circumference.

"When I first did this," Henes told *The New Yorker,* "I thought you had to use organic eggs. But it turned out you don't." She said she had no idea why eggs balanced on the equinox. "They just do, is all. I've had friends tell me you can even use eggs right out of the fridge. They don't even have to be at room temperature."

Several hundred peaceniks turned out for the 1983 ritual. Music was provided, said *The New Yorker,* by "two ocarinas, two saxophones, one sleigh bell, one harmonica, four tin whistles, and one tambourine." Peace messages written on several hundred orange streamers were tied to the iron railings surrounding the park. They bore such slogans as "World Friendship! Let's Have It Now!"; "The Universe Spreads Out Before Us, Ineffably Profound"; and "If Peace Comes to the Earth, Donna Will Be Largely Responsible."

For several minutes before the equinox, Henes chanted a peace slogan, then carefully balanced an egg on the concrete base of an abstract sculpture called "Peace Form One." All over the small park, eggs were balanced on the pavement, even on the iron railings. One man balanced an egg on the First Avenue median strip, where it stayed until it was bashed by a Checker cab. Henes moved through the crowd, rubber-stamping eggs with "This Egg Stood Up, 3/20/83."

The *New Yorker* reporter was impressed. None of the physicists contacted by the magazine had heard about equinox egg balancing, nor could they think of any reason why they would balance. Magician James Randi told the magazine that eggs balanced just as easily on any other day, but the *New Yorker* reporter didn't buy it. Two days later the reporter took a dozen eggs to Ralph J. Bunche Park and for twenty minutes was unable to balance a single egg.

Such self-deception is not hard to understand. If you are convinced that an egg will balance more easily on a certain day you will try a little harder, be more patient, and use steadier hands. If you believe that eggs won't balance on other days, this belief is transmitted subconsciously to your hands. It's the old Ouija-board phenomenon.

Even *The New Yorker* admitted this possibility:

The trouble may have been that we didn't want the egg to balance—that we wished to see Donna Henes proved right. Something she had said to us shortly after the equinox kept running through our mind. "When I hold an egg at just that moment," she had said, "I feel as if the whole universe were in the palm of my hand. And when it balances, when it stands there, it's very calming. I feel so protected. It's as if the whole universe were working fine."

Whether an egg will balance or not depends on many conditions other than steady hands. The main factors are roughness of the egg's end and the roughness of the surface upon which the egg is placed. A concrete surface, for example, is so extremely uneven that it is not difficult to find a spot where any egg will balance. Moreover, because of slight surface irregularities on the eggshell itself, it sometimes will balance even on a smooth tabletop. If, however, you sandpaper the egg's end until it is perfectly smooth, balancing it on glass or Formica is impossible.

Henes's annual egg-balancing ceremony continued for many more years. In 1984 five thousand people participated in the event when it was held at the plaza of the World Trade Center.

Scot Morris, in his monthly column in *Omni* (March 1987), covered Henes's tenth annual ritual. "I don't know why it works," Henes told Morris, "but it does. Maybe it's because for a time surrounding the exact moment of the equinox the sun is directly over the equator and the earth is balanced within the universe."

No one asked Henes why it works so well in China on February 4 and 5. Astrology buffs have a similar difficulty explaining why astrology works so well in China and India, where it bears no resemblance to Western astrology. All three astrologies can't be right!

Some believers claim that eggs also balance easily on the autumnal equinox, about September 23, but the vernal equinox continues to be the most popular date. I do not know if this is still an annual event in Manhattan. The most recent report about it I could find in the *New York Times* was in 1988. An editorial of March 19 was headed, "It's Spring, Go Balance an Egg." Next day's *Times* said scores of people planned to gather on March 21 at the World Trade Center to start standing eggs on end at precisely the start of the equinox. A photograph of the event ran in the *Times*'s March 21 issue. Robert Novick, a Columbia University physicist, is quoted as saying that gravitational forces are far too weak to have any influence on the eggs. I am told by Morris that Henes has moved to San Francisco. My letter to her was not answered.

Magicians have a way of balancing eggs on hard, white surfaces by cheating. Make a tiny pile of salt. Balance an egg on the pile—you can even use its sharp end—then gently blow away the salt. A few undetectable grains remain to keep the egg upright.

The story of how Christopher Columbus balanced an egg was first told by Girolamo Benzoni in his 1565 *History of the New World*. Columbus was said to have attended a party where someone told him that even if he had not found the Indies someone else from Spain subsequently would have. Columbus asked for an egg. He challenged those present to balance it. After they failed, he balanced the egg by crushing an end. His point was that once a deed is done, it is easy to see how to do it.

Fifteen years earlier a similar story had been told by Giorgio Vasari in his 1550 *Lives of the Most Eminent Painters, Sculptors, and Architects*. The Italian architect Filippo Brunelleschi had designed a dome for a cathedral in Florence named Santa Maria del Fiore. The city fathers demanded to see his model, but he refused. Instead, he challenged a group of architects to make an egg stand on end. Whoever succeeded, he told them, would be allowed to build the dome. After they all failed, he demonstrated how it could be done by tapping the egg on a marble table to flatten one end. "The craftsmen protested that they could have done the same, but Filippo answered laughing that they could also have raised the cupola if they had seen his model. And so it was resolved that he should be commissioned to carry out this work." When the church was finally built, years before Columbus made his voyage, it had the shape of half an egg slightly flattened at the top.

A popular type of mechanical puzzle is an egg that can be balanced only if you discover its secret. Jerry Slocum, of Beverly Hills, California, who owns the world's largest collection of mechanical puzzles, provided me

with a history of balancing eggs. He sent seventeen pages from old catalogs advertising such eggs, beginning with Montgomery Ward's "Columbus Egg" of 1894. He also sent the first pages of eighteen U.S. patents,
starting in 1891, for balancing eggs. Their internal mechanisms vary widely.
They include weights to be manipulated, mercury to be maneuvered
through tubes, and steel balls to be rolled up spiral paths to the egg's center or guided through a maze.

"Professor Hoffmann," in *Puzzles Old and New* (London, 1893), describes a Columbus Egg containing a hollow cone with a hole at the top.
The puzzle is solved by rolling a ball up a groove until it drops into the
cone and falls to its base. You'll find a color photo of this egg in L. E.
Hordern's privately published edition of Hoffmann's book (London, 1993),
richly illustrated with photographs of puzzles in Hordern's collection. Puzzle collector Robert Darling, of Johnson City, Tennessee, gave me an ingenious egg currently sold in Germany by a firm named Pussycat. It
balances only if you hold it with its pointed end upright for twenty-five
seconds, then quickly invert it. It will then balance on its pointed end for
fifteen seconds before it topples over.

Finally, I must mention Piet Hein's celebrated superegg, discussed in
Chapter 18 of my book *Mathematical Carnival* (1977, Random House).
It legitimately balances on either end without any trickery.

Addendum

Monty Vierra wrote from Taiwan to say that on Taiwan egg balancing takes
place not on the first day of spring but on the fifth day of the fifth lunar
month, called *duan wu jie*. It occurs sometime in the Western world's June.

Donna Henes, who bills herself as an "urban shaman and ceremonial-
ist," continues to sponsor egg-balancing events in Manhattan every spring.
She is also tirelessly active in New Age circles from her Mamma Donna's
Tea Garden and Healing Haven, PO Box 380403, Exotic Brooklyn, NY
11238-0403. The shop sells a variety of "multicultural ritual tools and cer-
emonial supplies including specially blended teas." For $75 an hour any-

one can have a private consultation with Mamma Donna, with a two-hour minimum for the first appointment.

Henes's book *Celestially Auspicious Occasions: Seasons, Cycles and Celebrations* was published by Perigee in 1996. In 1998 she began issuing a quarterly newsletter titled *Always in Season: Living in Sync with the Cycles.* In February 1998, Henes sponsored an expedition to the West Indies island of Antigua to view the last total solar eclipse visible this century in the western hemisphere. The fee was $729 per person. According to an announcement I received, there would be drumming, dancing, and chanting on the beach. Astrologer Geraldine Hannon was listed as available to offer information, and psychic Patricia Einstein would be there to talk about the effects of eclipses on "the creative unconscious, myth, and symbol."

An 1892 Columbus Egg from Jerry Slocum's collection. The tiny lead ball must be guided through the open tube to the base, where it stabilizes the egg.

Balancing eggs from Jerry Slocum's collection.

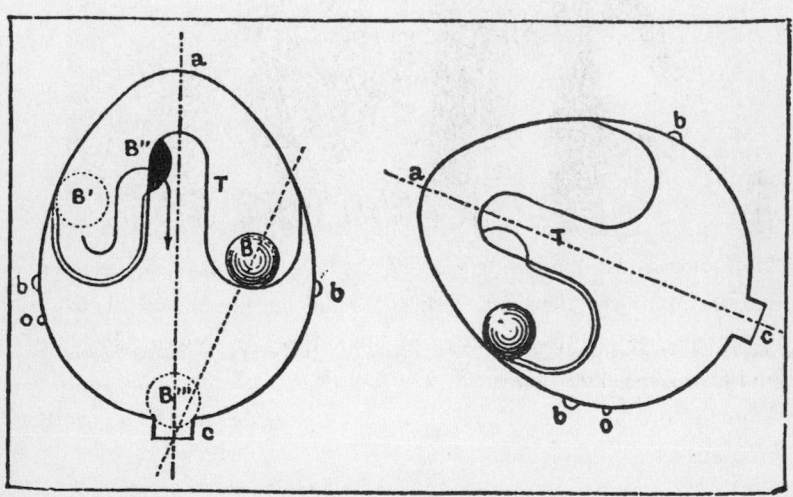

Details on the construction of a balancing egg sold in London. From Albert Hopkins's *Magic, Stage Illusions and Scientific Diversions Including Trick Photography* (Munn, 1898). Hopkins recommends covering the egg with a handkerchief while you make it stand on end so that spectators will not see how the trick is done.

CHAPTER 6

Zero-Point Energy and Harold Puthoff

> I always get the creeps when people talk about virtual particles.
> —*Victor Weisskopf, as quoted by K. C. Cole*
> *in* Science as Metaphor

> One longs for a new Einstein who will, in a flash of insight, give
> us back our lovely nothingness.
> —*Leon Lederman, in* The God Particle *(1993)*

I n the December 1997 issue of *Scientific American,* staff writer Philip Yam's article "Exploiting Zero-Point Energy" is devoted to a ten-year struggle by physicist Harold E. Puthoff to build a device that could tap the fluctuating energy of supposedly empty space-time. An episode called "Beyond Science" of PBS's *Scientific American Frontiers,* which aired on television the previous month, also had a segment devoted to Puthoff's ambitious research program.

What *Scientific American* failed to reveal, both in Yam's excellent piece and on its TV show, was that Puthoff is none other than the Harold Puthoff who twenty years ago validated the psychic powers of Uri Geller. In 1976 Puthoff and his friend Russell Targ were on the staff of what was then called the Stanford Research Institute (SRI), now SRI International. Their book *Mind-Reach* (1976) tried to convince the world that ESP, PK (psychokinesis), and precognition now have, thanks to their valiant efforts, become firmly established phenomena. Margaret Mead wrote the book's enthusiastic introduction.

Most of the work of Puthoff and Targ at SRI was devoted to what they called "remote viewing"—the ability of psychics to "see" scenery at any distance away—perhaps even to remote-view the surfaces of other planets. Chapter 7 described experiments which they said proved that Israeli magician Uri Geller had strong psychic powers. In later papers Puthoff and Targ claimed astonishing success with an ESP teaching machine. They also claimed to have validated Geller's ability to guess correctly how a die had fallen when shaken inside a closed box.

The original manuscript of *Mind-Reach* contained several pages outlining what the authors insisted was a surefire technique of using precognition to win large sums of money at roulette tables. Although Mead believed strongly in paranormal powers, she objected so vigorously to including this betting method in the book that it was removed from the published edition, though not from proofs sent to reviewers.

Prior to his work at SRI, Puthoff was an active Scientologist. He had been declared what the group calls a "clear"—a person free of "engrams." Engrams are alleged to be records on an embryo's brain, long before it grows ears, of what its pregnant mother is speaking or hearing. These records are said to cause neuroses and psychoses in one's adult life. When Puthoff married, a Scientology minister performed the ceremony. The Church of Scientology proudly published a 1970 notarized letter written by Puthoff when he was a Stanford University physicist specializing in laser research, a topic on which he had coauthored a textbook. Five years earlier he had earned his doctorate in electrical engineering at Stanford.

"Although critics viewing the system [Scientology] from the outside," Puthoff wrote in his letter, "may form the impression that Scientology is just another of many quasi-educational quasi-religious 'schemes,' it is in fact a highly sophistical and highly technological system more characteristic of the best of modern corporate planning and applied technology."

The letter goes on to praise Scientology's E-meter, a simple electronic device used by "auditors" to uncover a patient's engrams. "In the technical community here at Stanford, we have projects underway employing the techniques developed in Scientology." Puthoff adds that Scientology is an "uplifting and workable system of concepts which blend the best of Eastern and Western traditions. After seeing these techniques in operation and experiencing them myself, I am certain that they will be incorporated eventually on a large scale in modern society as the readiness and awareness level develops."

L. Ron Hubbard, the science fiction writer who invented Scientology and became its guru, wrote a book titled *Scientology: A Religion.* Puthoff provided its preface. In it he blasts the FDA for calling the E-meter useless. He likens attacks on Scientology to attacks made on Harvey, Galileo, Semmelweiss, and Copernicus. "Nevertheless," he concludes, "it is incumbent upon the pioneers of new developments to press forward their discoveries in the face of all opposition."

After leaving SRI in 1987, Puthoff was hired by a think tank in Austin, Texas, called the Institute for Advanced Studies. (It has no connection with the institute of a similar name in Princeton, New Jersey.) On May 28, 1987, at an Austin conference, Puthoff gave a speech on "One Hundred Years of Remote Viewing." He praised the value of precognition in making stock market predictions and the ability of remote viewers to detect astronomical features of planets before those features are seen by telescopes or space probes. This was followed by a statement of which I hope Puthoff is now thoroughly ashamed. He referred to a thirty-eight-year effort by two followers of Madame H. P. Blavatsky, founder of theosophy, to remote-view the inner structures of atoms!

Annie Besant and C. M. Leadbeater published their curious results in a book titled *Occult Chemistry* (1908). It swarms with drawings of the inner structures of atoms. These outlandish sketches have absolutely no scientific merit, but Puthoff was convinced that occasionally they anticipated modern particle physics! In an excerpt from his speech, quoted in *The Explorer* (Vol. 4, October 1987), Puthoff called these sketches a "remarkable" study of the "basic constituents of matter." The sketches, he said, "found relatively little correlation with known scientific fact until the recent development of quark and superstring theories, which show striking correspondence to the reported observations." The striking correspondence, alas, is visible only to Puthoff and theosophists.[1]

I do not know what Puthoff now believes about the Besant-Leadbeater micro–remote viewing of the interior of atoms, or about the "genius" of L. Ron Hubbard, or the efficiency of E-meters. Reincarnation was one of a raft of strange doctrines Hubbard added to Dianetics when he turned it into a tax-free religion. Today it is as essential a belief for Scientologists as it is for theosophists. Puthoff is on record as saying he no longer is associated with Scientology, but how much of it does he still buy? Does Dr. Puthoff still think that mental ills can result from experiences in previous lives? As far as I know, Puthoff no longer conducts experiments in remote viewing. He and Targ went their separate ways after leaving SRI.

For the past decade, Puthoff's tireless efforts at the Institute for Advanced Studies, where he is now director, have gone into searching for a way to obtain unlimited free energy from the quantum fluctuations of empty space. To almost all other experts, this search is as quixotic and futile as the search for a perpetual motion machine. They see the situation as comparable to having research on how the brain works directed by a neuroscientist who believes in phrenology. According to Yam, Puthoff's institute has examined about ten devices for tapping the energy of space, all of them failures.

Zero-point energy (ZPE) is a term for the energy that constantly fluctuates in the vacuum of space and at the heart of all matter. If the temperature of matter could be lowered to absolute zero, it was once thought, its atoms and inner electrons would stop moving and the matter would collapse. It is now known this cannot occur. ZPE keeps the atom constantly

---

[1]For details on the remote viewing of atoms, consult *Extrasensory Perception of Quarks* (Theosophical Publishing House, 1980), by British physicist Stephen M. Phillips; his two-part article "Extrasensory Perception of Subatomic Particles," in *Fate* (April and May 1987); and "Resolution in Remote-Viewing Studies: Mini- and Micro Targets," by Puthoff, Targ, and Charles Tart, an SRI report of June 1979.

jiggling. Heisenberg's famous uncertainty principle forbids it to become motionless.

This jiggling also applies to any particle supposedly at "rest." Imagine an electron being squeezed into a smaller and smaller space by a piston. As the electron's position becomes more accurately known, the uncertainty relation ensures that its momentum becomes fuzzy and more intense. The electron cannot be totally motionless because then its position and its zero momentum would be precisely known. As the electron is squeezed into an increasingly tiny space, its pressure on the piston increases as it strikes the piston with greater force and frequency. It is this pressure of electrons within every atom that preserves the atom in what is called its "ground state."

The incessant fluttering of all particles when at absolute zero has been verified in numerous ways. The Lamb shift, for example, results from the action of ZPE on spectra. In the famous Casimir effect, ZPE forces two parallel metal plates to move closer together. ZPE causes low-level noise in microwave receivers. It excites the atoms in fluorescent lamps. It plays a role in the surface tension of liquids, in images on eye retinas, in the scattering of light that makes the sky blue, and in many other physical phenomena. In cosmology it sends out radiation from black holes. Its pressure prevents gravity from collapsing white dwarf stars.

Heisenberg's uncertainty principle also underlies one of the most bizarre aspects of quantum theory. The vacuum of space-time is by no means "nothing." It is a foaming sea of constantly bubbling particles that flash into existence for fleeting microseconds only to be absorbed back into the mother sea from which they momentarily borrowed a tiny bit of energy.

Time and energy, like position and momentum, also are subject to the uncertainty relation. If the time during which energy is measured is known exactly, the amount of energy becomes uncertain. The shorter the time interval, the greater the uncertainty. When the interval is short enough, it allows energy to appear from nowhere in the vacuum of space provided it vanishes fast enough back into the mother sea to preserve the vacuum's overall zero energy.

This energy that randomly pops out of empty space takes the form of particle-antiparticle pairs that mutually annihilate. This happens much too fast to be observed, but can be inferred from other phenomena. On the average, the pairs exist for about 0.00000 00000 00000 00000 1 of a second, with a maximum distance between them of about 0.00000 00001 of a centimeter.

Every type of particle known is believed to emerge briefly from the churning vacuum, the lighter particles such as electrons and photons more frequently than heavier particles such as protons, neutrons, and quarks. It is theoretically possible that a macro object such as an apple might be created for an instant, but the probability of this is far too low to allow it. These ghostly particles are called "virtual" to distinguish them from their "real" forms that persist in time.

The fluctuation of particle pairs occurs within all quantum fields, but mainly in electromagnetic and gravity fields. The gravity field presumably generates the conjectured, but so far undetected, massless graviton-antigraviton pairs. The energy-time uncertainty also allows every real particle to be surrounded by a cloud of virtual particles of all varieties that are constantly being emitted and absorbed by the seething vacuum that surrounds the real particle.

Here is how Heinz Pagels, in *The Cosmic Code,* describes the vacuum of space:

> Space looks empty only because this great creation and destruction of all the quanta takes place over such short times and distances. Over long distances the vacuum appears placid and smooth—like the ocean which appears quite smooth when we fly high above it in a jet airplane. But at the surface of the ocean, close up to it in a small boat, the sea can be high and fluctuating with great waves. Similarly, the vacuum fluctuates with the creation and destruction of quanta if we look closely at it.

In 1973, physicist Edward Tryon made a startling proposal in a two-page paper titled "Is the Universe a Vacuum Fluctuation?" (*Nature,* Vol. 246, pp. 396–97). He suggested that a vacuum fluctuation may have triggered the big bang! As he put it, "Our universe is simply one of those things which happen from time to time." This implies that space and time existed before the bang. Other physicists have since proposed slightly different ways a quantum fluctuation in a vacuum devoid of space and time could create a runaway universe, though how something could fluctuate without space and time is unclear. Of course our universe could not emerge from absolutely nothing. There would have to be quantum fields to fluctuate, leaving unanswered the ultimate question of where quantum fields and their laws came from, or why there is something rather than nothing.

In recent years a number of physicists have wondered if it is possible to somehow tap the ZPE of the fidgety vacuum. Most physicists consider this hopeless. At the end of the PBS's *Scientific American Frontiers* show, Steven Weinberg, the Nobel Prize–winning physicist now at the University of Texas, in Austin, pointed out how weak this energy is. In the entire universe, he said, it is enormous, but the total amount of ZPE available in a space the size of the earth is about the same as the energy obtainable from a gallon of gasoline. Trapping this energy of course means that a machine must be able to snatch energy from the virtual particles before they disappear. No one has any good idea of how this could be done, and even if it could be, the energy available would be insignificant.

British physicist Paul Davies, in *Other Worlds* (1980), had this to say in Chapter 4: "There is no question . . . of running a machine on borrowed energy. . . . The energy output from an electric light emitted in one second can only be borrowed via the uncertainty principle for a billion-billion-billion-billionth of a second. Put another way, the quantum loan mechanism can only enhance the output from an electric light by one part in one followed by 36 zeros."

Puthoff disagrees. Like other mavericks working on ZPE machines, he sees the opposition of mainstream physicists to such research as the irrational knee jerks of an elite. "Most working physicists are not really scientists," he told an interviewer in 1990. "They are number crunchers, computer operators, lab technicians. It's not all their fault. It's driven largely by the military-industrial complex."[2] In many technical papers and several

[2]Puthoff is so quoted in "Power Structure," by Tom Chalkey, in Baltimore's *City Paper,* June 29, 1990.

popularly written articles he has defended the possibility of obtaining unlimited energy from empty space. In *Scientific American*'s PBS broadcast he predicted that just as this century is known as the nuclear age, so will the next millennium be known as the zero-point energy age.

Puthoff sees himself as a lonely pioneer whose research he is confident will usher in this awesome new age. In his paper "Quantum Fluctuations of Empty Space: A New Rosetta Stone of Physics?" (a speech reprinted in *Frontier Perspectives*, Vol. 2, Fall/Winter 1991, pp. 19–23, 43) he predicts that tapping ZPE will revolutionize history. "Only the future," he concludes, "will reveal to what use humanity will eventually put this remaining fire of the gods. . . ."

Many of Puthoff's recent conjectures are far out on the fringes of physics. He believes that gravity may be caused by ZPE in a manner similar to the way it causes the Casimir effect. He suggests that electrons are kept in their atomic orbits by ZPE, and that if atoms could be "shrunk" to a lower ground state they would radiate ZPE. Inertia, he thinks, may be caused by resistance of ZPE whenever objects are accelerated. If this resistance could be reduced, it would provide a great advance in the rocket propulsion speed of spaceships. In "SETI, the Velocity-of-Light Limitation, and the Alcubierre Warp Drive: An Integrating Overview" (in *Physics Essays*, Vol. 9, 1996, pp. 156–58), Puthoff defends the possibility that spaceships could travel faster than light if the ZPE could be handled properly.[3]

Just as Puthoff and Targ were able to obtain millions in funding dollars for their SRI research on remote viewing, so Puthoff is now raking in funds from sources he prefers not to disclose. It remains to be seen if in the next few decades this eccentric physicist will turn out to be one of the greatest scientists of all time, or whether his ZPE speculations and work will blow away like the flawed research he supervised when he and Targ were in their glory days at SRI.

---

[3] Arthur C. Clarke, in his 1997 novel *3001*, takes Puthoff's conjecture about inertia seriously enough to base an inertial space drive on ZPE. See Chapter 9 of the novel.

Addendum

Puthoff's lengthy reply to my column was published in the September/October 1998 issue of *Skeptical Inquirer*. He vigorously defended the value of his research, stressed that he was no longer active in Scientology, but said nothing about his beliefs in the paranormal. Here is how I replied:

> I'm delighted to learn that Harold Puthoff has severed his ties with the Church of Scientology, though to what extent he may still accept some of its basic doctrines remains unclear. In any case, to have once taken L. Ron Hubbard seriously is no tribute to Puthoff's youthful sagacity.
>
> The most striking aspect of Puthoff's letter is what it does *not* say. There are no hints concerning his present opinions about ESP, PK, and precognition. Nor does he repudiate his conviction that two theosophists, more than a century ago, remote-viewed the interior of atoms with results unexplainable by chance.
>
> Regarding ZPE (zero-point energy), Puthoff lists three physicists as "laying the groundwork" for his efforts to obtain unlimited energy from the vacuum of space. Andrei Sakharov is cited only because he once speculated that gravity might be the result of ZPE when altered by matter—a wild conjecture that has led nowhere. I can't imagine why Paul Davies is mentioned. Did Puthoff forget that I quoted Davies as saying it was folly to search for a ZPE machine?
>
> Robert Lull Forward is the only physicist among the three who reasonably can be said to have furnished "groundwork" for Puthoff's research. Forward is a maverick physicist, best known for his "hard" science fiction novels such as *Dragon's Egg, The Flight of the Dragon Fly, Martian Rainbow,* and *Timemaster.* His nonfiction *Future Magic,* a 1988 Avon paperback, does indeed contain suggestions about how ZPE machines might be constructed. Forward also describes devices for producing repulsive gravity, for traveling backward in time, and other fantasies. Avon summarizes the book as identifying the boundaries of the human soul, flying to stars on beams of light, climbing a magic beanstalk into space, soaring through the solar system on magic matter, and defying gravity.

Forward ends his book by predicting that quantum mechanics will someday provide a "natural" explanation of ESP that will remove it "from the pages of the tabloids" and place it on "the pages of scientific journals." He defends the view that our brain's pattern of molecules constitutes an immortal soul. "It may be," he concludes his book, "that on some future-magic day, rather than denying the existence of the spirit, science will prove that the spirit *does* have a physical reality and that there *is* life after death."

It goes without saying that research on the nature of ZPE should continue, but that was not the point of my column. The point was: Is Puthoff, with his record of psychic research, and his beliefs about the paranormal, a good man to trust with large funds for efforts to build what most physicists think is apparatus as incapable of providing free energy as a perpetual motion machine?

Let me repeat the question I asked in my column: "Puthoff: do you still believe that [Uri] Geller is a genuine psychic with remarkable paranormal powers?"

I'm still waiting for an answer.

Physicist Steven Shore, an editor of *The Astrophysical Journal,* wrote to *Skeptical Inquirer.* He said that Puthoff's *Physical Review* papers may have gone unanswered, but his submissions to *The Astrophysical Journal* were routinely rejected on the grounds that his research is flawed and trivial.

Phil Klass sent me a copy of Puthoff's review of a book by Paul Hill titled *Unconventional Flying Objects,* a 1995 work speculating on the technology behind UFO propulsion systems. Puthoff praises the book as the "most reliable, concise summary" of data available from UFO sightings. The review appeared in the *Journal of Scientific Exploration* (Vol. 10, No. 4, 1996), published by Peter Sturrock's Society for Scientific Exploration. The society, of which Puthoff is a member, publishes papers about ufology and other dubious fringe sciences.

For Puthoff's most recent paper on using zero-point energy for the propulsion of spaceships, see "Can the Vacuum Be Engineered for Spaceflight Applications? Overview of Theory and Experiments," in Sturrock's magazine (Vol. 12, No. 2, 1998). Puthoff concludes by quoting Arthur

Clarke's aphorism that advanced technology is indistinguishable from magic. "Fortunately," Puthoff ends his paper, "such magic appears to be waiting in the wings of our deeper understanding of the quantum vacuum in which we live."

Arthur Clarke continues to be optimistic about the possibility of tapping zero-point energy. Clarke was interviewed in *Free Inquiry* (Spring 1999). Asked what he thought about chances for a great new energy revolution, he replied:

> I don't know whether it will come in cold fusion or warm fission or something else. I suspect it might be something totally unexpected—perhaps a way of tapping into quantum fluctuations of space—zero-point energy, as it's sometimes called. Now, this new finding may turn out to be an experimental laboratory curiosity that can't be scaled up. But remember, nuclear power started as a small laboratory curiosity.

The Casimir effect occurs when thin metal plates are extremely close together. Virtual particles of longer wave lengths, excluded from between the plates, generate enough force to push the plates together. In 1999 electrical engineer Jordan Maclay obtained a NASA grant to work on a Casimir machine for tapping ZPE. See Henry Bortman, "Energy Unlimited," in *New Scientist,* January 22, 2000, pp. 32–34.

Physicist Victor Stenger, in an article titled "The Phantom of Free Energy," in *Skeptical Briefs* (June 1999), takes a dim view of Puthoff's work. He calculates that if the Casimir effect is used for extracting energy from the vacuum, it would require two squares, each two hundred kilometers on the side and separated by a millionth of a meter, to light a hundred-watt bulb for one second. "If we were to stumble upon thirty million or so of these structures out in space, we could hook them up to our lightbulb and keep it lit for a year." Stenger closes: "I do not recommend that you invest your retirement funds in any companies that promise to develop this technology."

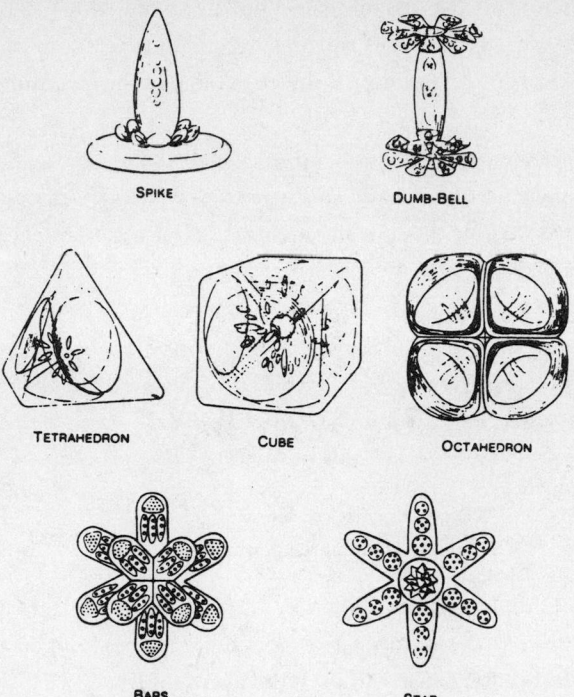

SPIKE

DUMB-BELL

TETRAHEDRON

CUBE

OCTAHEDRON

BARS

STAR

Some typical examples of atomic structure as "seen" clairvoyantly by two leaders of Theosophy and published in their 1910 book *Occult Chemistry*.

CHAPTER 7

David Bohm

The Guided Wave

> When to the new eyes of thee
> All things by immortal power,
> Near or far,
> Hiddenly
> To each other linked are,
> That thou canst not stir a flower
> Without troubling of a star.
> —*Francis Thompson*, The Mistress of Vision

> I am working out a quantum theory about it for it is really most
> tantumising state of affairs.
> —*James Joyce*, Finnegans Wake

There is little doubt that the mathematical formalism of QM (quantum mechanics) is accurate. No theory of physics has been more spectacularly successful in making predictions about the outcome of measurments. Some are accurate to many decimal places. Where experts disagree is not about the mathematics of QM but about how to interpret its equations. Even more than relativity theory, QM bristles with wild paradoxes that radically violate common sense, and for which at present there are no agreed-upon resolutions.

The most notorious of these paradoxes is the EPR, named after the initials of Einstein and two associates, Boris Podolsky and Nathan Rosen. In 1935 they published an explosive paper in which they argued that their paradox proved that QM is incomplete, destined to be replaced or radically modified by a deeper theory.

The EPR paradox has several forms, but the easiest to understand was proposed by the late American physicist David Jacob Bohm. (Note how his last name differs from Bohr's by only one letter.) It involves a mysterious property of particles called spin. Spin is roughly similar to the spin of a top, because it has angular momentum that always takes one of two forms variously called left or right, plus or minus, up or down. Imagine a quantum reaction that creates two identical particles A and B which go off in opposite directions. In standard QM each particle has its left and right spins "superposed." When particle A is measured for spin, its "wave function" (a formula specifying the probabilities that certain values will be found when a particle is measured for a given property) is said to "collapse" (vanish). The particle at once acquires either a left or right spin with equal probability.

Now for the magic. To conserve angular momentum, after A is measured and so acquires a definite spin, B must acquire the opposite spin. Assume that A, measured in Chicago, has a left spin. (Remember, it does not have a definite spin until measured.) On a planet in a distant galaxy a physicist measures B when it gets there. It is certain to have a right spin. How does B "know" the outcome of the measurement of A? Does A send some kind of telepathic signal to B, either simultaneously or at a speed equal to or exceeding the speed of light? Einstein ridiculed this as "spooky action at a distance." He believed that his proposed experiment, then only a thought experiment, proved that QM was not complete. There must be local "hidden variables" giving definite spins to both particles befor one is measured.

The standard Copenhagen interpretation of QM, based on the opinions of Niels Bohr, is that regardless of how far apart A and B get, they remain a single quantum system with a single wave function. When A is measured, the entire system's wave function vanishes and the two particles simultaneously acquire opposite spins. The particles are said to be "correlated," or in more recent terminology, "entangled."

Does this resolve the paradox? It does not. The mystery remains of how A and B can stay entangled when they are light-years apart unless there is some kind of connection between them that allows information to go from A to B.

All physicists agree that there is no possibility of sending coded messages by using the EPR phenomenon. The situation is like two persons, one in New York, the other in Paris, who simultaneously flip a penny. For reasons unknown, if one penny falls heads, the other must fall tails, and vice versa. If one could control the outcome of a flip in Chicago, then of course the Chicagoan could send a message in a binary code of ones and zeros. But there is no way to control the outcome of a measured spin. Like the heads and tails of a flipped coin, it becomes left or right with equal probability. The sequence of lefts and rights, at each end, is always a random, meaningless sequence. If it were otherwise, cipher messages presumably could be sent with a speed exceeding that of light, thereby violating a basic law of relativity.

The EPR paradox remained unconfirmed until recent years when the late Irish physicist John Stewart Bell thought of a brilliant way to test it over short laboratory distances. The paradox has since been completely validated many times. The two entangled particles behave precisely as QM predicts.

Several conflicting efforts have been made to explain the EPR paradox. Here I confine my remarks to how it is resolved by what is called the pilot-wave or guided-wave theory (GWT). Obviously I'm no expert on QM, only a science journalist, so I haven't the foggiest notion of whether GWT will someday be confirmed. However, this theory, long ignored by physicists, is now gaining increasing support.[1] It deserves to be better known.

---

[1] See, for example, David Z. Albert's vigorous defense of Bohm's theory in "Bohm's Alternative to Quantum Mechanics," in *Scientific American* (May 1994). Albert is a professor of philosophy at Columbia University, with a Ph.D. in physics. More on Bohm's theory can be found in his book *Quantum Mechanics and Experience* (Harvard University Press, 1992).

It was Louis de Broglie, one of the architects of QM, who first proposed GWT. He early abandoned it after heavy ridicule by Copenhagenists, but took it up again after it was improved by Bohm. Now known as the Broglie/Bohm GWT, it has become the interpretation of QM favored by a raft of experts that include Bell, Jeffery Bub, the French physicist Jean-Pierre Vigier, and many others. So far its predictions are exactly the same as those of the Copenhagen school, although there may be subtle ways it can be tested by difficult experiments not yet performed.

In standard QM, every particle can be observed either as a particle or as a wave. The wave is not physical, like water or sound waves, but a wave of probability in an abstract space. When a photon goes through one slit in a barrier, to register on a detection screen, it is a particle. When two slits are open, the photon behaves like a wave and it is impossible to tell which slit it goes through without destroying the wave. If many photons are sent through a barrier with two openings, each registers on the screen as a particle, but they display an interference pattern that could only be produced by a wave going through both slits. The photon is a mysterious thing. It is neither wave nor particle, but something that can act like one or the other depending on the measuring apparatus.

In Bohm's revolutionary theory, as refined by his associate Basil Hiley, particles are as real as golf balls. At all times they have precise, unfuzzy properties such as position and momentum, and precise paths through space-time. The particles are *never* waves. Associated with each is an invisible, undetectable wave in a field which Bohm called the "quantum potential." Its pilot waves are real waves, not probability waves. They guide the particle's motion in a manner somewhat like the way a river's wave guides the movement of a floating leaf, or, in a better analogy, the way radar information guides a ship. This quantum field, like the fields of gravity and electromagnetism, permeates all of space-time, but unlike those fields its intensity doesn't diminish with distance. Also unlike other fields, it exerts no force on particles. Essentially it is a wave of undecaying information.

It is the *ad hoc* nature of this undetectable pilot wave that reminds so many of Bohm's antagonists of the old stagnant aether of the nineteenth century, a substance that could not be detected as a carrier of electromagnetic waves, and that Einstein had discarded as useless. As J. C. Polking-horne, in his marvelous little book *The Quantum World* (1984), said: "In the opinion of many Bohm had jumped out of the indeterminate frying pan into a crackling nonlocal fire."

When just one slit is open, the pilot wave guides each photon through the opening and there are no interference bands. When both slits are open, each particle goes through just one slit, but its pilot wave, quite separate from the particle, goes through both to guide a stream of photons along paths that produce the wave interference pattern on the screen. It is not a case of an entity being either a wave *or* a particle, but a case of there always being present both a wave *and* a particle. Comic versifier Armand T. Ringer put it this way:

> As a photon flew close to slit 2,
> She said to her pilot wave, "You
> Must slide through both slits,
> Or Bohm's fans will have fits,
> And be forced to abandon their view."

Here is how GWT explains the EPR paradox. When A is measured, its guiding wave, in the unobservable quantum potential field, twists A into either a left or right spin. At the same instant, no matter how far away B is, it twists B the opposite way. This unobservable field, extending through-out the cosmos, has the nonlocal power to act simultaneously on both par-ticles. No information travels from A to B. Instead, it travels simultaneously from the quantum field to the two particles, giving them opposite spins. This, of course, is action at a distance, and was probably the main reason Einstein did not care much for Bohm's theory.

How does the pilot wave manage to guide the paths of particles? This is one of the darkest mysteries of GWT. It is able to push particles around without at the same time exerting any force on them. If it did, photons would have their energies altered. But this doesn't happen. The photons ar-rive on the screen without any change of energy. Somehow each photon must pick up information from its pilot wave without having its energy modified. This may be spooky, but no spookier, say Bohm's supporters, than the probability waves in orthodox QM which decide on how a wave function will create certain properties when it collapses.

Could the two particles of the EPR paradox be just projections in our space of a single particle moving in a higher dimension? Bohm once speculated on this possibility. Imagine a tank, he wrote, in which a fish is swimming. Two television cameras film the tank from two different directions. We see only the two films projected on two screens. We think we are watching two fish, and are amazed by the curious way their movements are correlated. The tank is in the unseen higher dimension. In Bohm's later terminology, it is in the "implicate order" which lies beyond the "explicate order" of the world open to our experience. What we think are two fish are really projections in our world of a single entity.

Bohm suffered all his life from the contempt with which followers of Bohr viewed his GWT. Bohr called it "foolish." J. Robert Oppenheimer called it "juvenile deviationism" and advised physicists to ignore the theory. Even Einstein, who for a time admired de Broglie's work, decided that Bohm's GWT was "too cheap" a way to resolve QM paradoxes.

Bohr's antagonism toward Bohm was extreme. The German-born physicist Ernest J. Sternglass, who favors Bohm's pilot wave, tells in his book *Before the Big Bang* (1997) about a meeting with Bohr during which they discussed Bohm's theory. "It was embarrassing," Sternglass writes, "for me to see Bohr so emotionally upset about Bohm's work. . . . The vehemence of this otherwise mild-mannered and kindly man really surprised me. . . . it seemed to me that Bohr had almost the fanatical approach of a fundamentalist preacher, intensely concerned to save my soul from perdition."

Mathematician David Wick in his great book *The Infamous Boundary* (1995)—the title refers to the gulf between the micro world of QM and the macro world of relativity theory, cosmology, measuring apparatus, and you and me—bashes Richard Feynman for a similar opposition to Bohm. Feynman is often quoted, from his famous "red books" of lectures, to the effect that nobody understands how the double slit experiment works. "The question is," Feynman asks, "how does it really work? What machinery is actually producing this thing? Nobody knows any machinery.

Nobody can give you a deeper explanation of this phenomena than I have given; that is, a description of it."

"Very well," comments Wick, "what is wrong with the pilot wave?" Wick attributes Feynman's "overheated rhetoric" to his refusal, like that of von Neumann, Heisenberg, and other quantum experts, to take Bohm seriously.

The great Hungarian mathematician John von Neumann wrote a famous book about QM in which he thought he proved that QM could never be modified by introducing hidden variables. By this he meant local hidden variables—variables attached to particles. He failed to realize that QM could be modified by introducing nonlocal variables such as Bohm's quantum potential. It is a scandal that so few of today's physicists and their students realize that Bohm succeeded in doing exactly what von Neumann thought impossible. We can describe his achievement by altering some words in a well-known poem by Eddie Guest:

> Von Neumann said that it couldn't be done,
>> But Bohm with a chuckle replied,
> That maybe it couldn't, but he would be one
>> Who wouldn't say so till he tried.
> "A quantum potential? The field is essential!"
>> If David Bohm doubted he hid it.
> He started to sing as he tackled the thing
>> That couldn't be done, and he did it!

Bohm's quantum potential binds the entire universe together into what he liked to call a seamless "unbroken wholeness." Every particle in the universe is connected by the quantum potential to every other particle. He likened the cosmos to a hologram in which each point on the film carries information about the entire picture. Bohm's GWT, far more sophisticated than de Broglie's crude version, is a "holistic" vision in which all parts of the universe are joined to every other part. "Interconnectedness" was one of Bohm's favorite words. He saw the universe as resembling the unity of a living organism, a kind of pantheism not unlike Spinoza's—a pantheism Einstein himself favored.

Although Bohm's GWT is identical with standard QM in its predictions, its way of talking about quantum phenomena is entirely different. The randomness of the Copenhagen interpretation which so disturbed Einstein ("God does not play dice") is replaced by a strict determinism. There are no quantum jumps. No superpositions. No collapsing of wave functions. Indeed, there is not even a "measurement problem." The probabilities of QM, which seem to spring from pure chance, in Bohm's theory result from our ignorance of the true, highly complicated state of affairs. The universe is real, "out there," independent of you and me. Human consciousness is not essential, as von Neumann, Eugene Wigner, and others supposed, to collapse wave functions.

Like Einstein, Bertrand Russell, Karl Popper, and almost all philosophers and scientists, Bohm was a thoroughgoing realist in believing that the universe with all its laws is independent of human minds. The moon is "out there," regardless of whether it is observed by any creature, such as a mouse. Bohm would have been horrified by the social constructivism of today's postmodernists who see all science and even mathematics as cultural creations similar to art, music, and fashions in clothes.

John Bell, who died in 1990 at age sixty-two, became an enthusiastic defender of Bohm. Bell was convinced that Einstein was intellectually superior to Bohr, a deeper thinker who saw clearly the need for QM to rest on a subquantum level (Bohm's potential field) that would restore both realism and determinism. Here is how Bell puts it in his paper "Six Possible Worlds of Quantum Mechanics," reprinted in his entertaining book *Speakable and Unspeakable in Quantum Mechanics* (1987):

> Is it not clear from the smallness of the scintillation on the screen that we have to do with a particle? And is it not clear, from the diffraction and interference patterns, that the motion of the particle is directed by a wave? De Broglie showed in detail how the motion of a particle, passing through just one of two holes in a screen, could be influenced by waves propagating through both holes. And so influenced that the particle does not go where the waves cancel out, but is attracted to where they cooperate. This idea seems to me so natural and simple, to resolve

the wave-particle dilemma in such a clear and ordinary way, that it is a great mystery to me that it was so generally ignored.

Bohm's sad life, his early pro-Soviet sympathies, his depressions, and his strange friendship with India's philosopher and mystic Jiddu Krishnamurti, are detailed in my *Skeptical Inquirer* column of July/August 2000. The column appeared too late to be included in this collection.

Part IV

Medical

Matters

CHAPTER 8

Reflexology

To Stop a Toothache, Squeeze a Toe!

Public infatuation with alternative medicines of all varieties shows no sign of abating. Acupuncture, homeopathy, aromatherapy, herbal remedies, chelation, iridology, therapeutic touch, magnet therapy, psychic healing, and so on are gaining new converts every day. The tragedies occur, of course, when gullible sufferers rely solely on such remedies and avoid seeking mainstream help. It would be good if we had some statistical evidence about the frequency of deaths following reliance on pseudomedicines.

Reflexology, one of the most preposterous of old alternative therapies now being revived, is the topic of this chapter. What is reflexology? It is the art of relieving pain and other symptoms of every ailment known to humanity by rubbing and massaging "reflex points" on the feet. I decided to write about it after I picked up a glossy, oversize volume on reflexology in a mall bookstore. It was hard to believe that this book had found a reputable American publisher.

The Complete Illustrated Guide to Reflexology, by Inge Dougans, was published simultaneously in 1996 by Element Books in England and Barnes & Noble in the United States. It is richly illustrated with full-color pictures on every page. The back of the book has a glossary of technical terms, a bibliography of earlier books on reflexology and related topics, and two pages listing schools and centers of reflexology around the world. Ten entries for the United States range from the Foot Relief Awareness Association, in Mission Hills, California, to the Pennsylvania Reflexology Association, in Quakertown.

According to the book jacket and an inside page, Inge Dougans was born in Denmark, where she had her first training in reflexology. In 1983 she founded the International School of Reflexology and Meridian Therapy, headquartered in Johannesburg, South Africa, where she practices. Branches are listed in Brazil, Canada, France, Germany, Italy, Netherlands, Portugal, Sweden, England, and New Jersey. From these centers you can obtain reflexology literature, posters, and a videotape giving a step-by-step guide to the therapy. Dougans also sells what she calls Vacu-Flex Boots. These are felt boots connected to a pump that creates a vacuum around the foot and applies uniform suction over the foot and ankle. Why suction would work as well as or better than air pressure is not made clear.

The origin of reflexology is hazy, but Dougans offers several conjectures. One is that it goes back five thousand years to ancient China; another that it originated among the Incas, who passed it along to Native Americans. Dougans reproduces an ancient Egyptian picture showing a man massaging another man's foot. She thinks this proves that reflexology flourished in ancient Egypt. Evidence for all these theories is nonexistent. What *is* known is that reflexology was a spinoff from a more general therapy called zone therapy that became popular in Europe, Russia, and America in the late nineteenth century.

Zone Therapy (1917), by two physicians, Dr. William Fitzgerald, an ear, nose, and throat physician at St. Francis Hospital in Hartford, Connecticut, and his associate, Dr. Edwin Bowers, is the classic American treatise on zone therapy. The authors divide the body into ten vertical zones, five on each side, that run like telephone wires from the top of the head to the fingers and toes. Two years later another American, Dr. Joseph Shelby Riley, added two horizontal zones. Riley wrote four books on the therapy, starting with *Zone Therapy Simplified* (1919). These zone lines are, of course, as imaginary as the energy lines on acupuncture charts.

In acupuncture, needles are inserted at specific spots on the body. In acupressure, another ancient Chinese therapy—it's a mild form of acupuncture—only pressure is applied at the reflex points. There is little agreement among various schools of acupressure over the precise locations of these points. For a while, in the mid and late 1970s, there was a flurry of interest in acupressure mainly because anyone could apply it to oneself.

In 1976, Lippincott published Yukiko Irwin's book *Shiatzu,* the Japanese term for acupressure. The book rated a full-page ad in the *New York Times Book Review* (February 29, 1976). *Cosmopolitan* (August 1975) ran a full-page ad for an illustrated acupressure course. "It works a lot like acupuncture," the ad reads, "only it's so simple, safe, and easy. You learn within minutes what can bring a lifetime of relief." Fifty illnesses are cited, including hemorrhoids, prostate trouble, diabetes, high blood pressure, and constipation—all of which acupressure offers "serious help in healing." If you are seasick, all you need do to make it go away is press a spot on your wrist that the Chinese call the *nei guan* or P6.

In the early 1980s a New York acupressurist named D.S.J. Choy thought of using a strap to apply pressure on P6. It was only a short time until a British firm was selling what it called "sea bands." A plastic button on each band applied pressure to the wrist. In the United States, similar "seasickness bracelets" came on the market, distributed by a firm in Palm Beach, Florida. Tests of these devices were inconclusive. About half of ship passengers said they seemed to diminish their sickness. The other half reported no effects.

Zone therapy was an American version of acupressure, although its pressure points bear little resemblance to those in Chinese and Japanese acupressure. Instead of using needles or pressure with fingers and thumb, zone therapy used tight rubber bands and spring clothespins to apply pressure. They were fastened on the finger or toe that was joined by a zone to the area where there was pain or some other distress symptom. Zone therapy placed no special emphasis on the feet.

Modern reflexology is that aspect of zone therapy which focuses on the feet. Eunice Ingham, an American who died in 1974, is called by Dougans the "Mother of Modern Reflexology" because she was the first to realize that the foot and its toes have especially sensitive spots for applying pressure. Her research was privately published in two popular books, *Stories the Feet Can Tell Thru Reflexology* (revised, 1938) and its sequel, *Stories the Feet Have Told Thru Reflexology* (revised, 1951).

Correlations of parts of the foot with the rest of the body, Dougans writes, "are similar to correlations with spots on the iris of the eye, the ear, and the hands." However, "corresponding areas of the feet are easier to locate because they cover a larger area and are more specific, rendering them easier to work on."

In today's reflexology the zones of zone therapy are replaced by twelve "meridians" along which a mysterious form of energy the Chinese call *ch'i* flows. Dougans, following the acupuncturists, likes to call it "yin and yang energy." She relates it to energy from sunspots, which she maintains have a strong influence on health. Peaks of sunspot activity, she says, correlate with epidemics such as Europe's Black Death and major flu outbreaks.

Dougans also discusses the influence of Earth's magnetic field on the mentally ill. Because the moon affects tides and the human body is 75 percent water, Dougans reasons, it is easy to understand why a full moon affects "arson, kleptomania, destructive driving, homicidal alcoholism," and other forms of psychotic behavior. As readers of *Skeptical Inquirer* know, numerous careful studies have shown that full moons have no such effects, in spite of the contrary opinions of many nurses.

Colorful charts in Dougan's big book identify scores of pressure points along the toes and on the soles and sides of the feet. The big toe, for example, has spots that lie on meridians to the hypothalamus, brain, mastoid, spine, and pituitary gland. The second and third toes connect to the eyes. The third and little toes join the ear. However, the tips of all five toes are on meridians leading to the sinus and teeth: "incisors on the big toe, incisors and canine teeth on the second toes; premolars on the third toes; molars on the fourth toe; wisdom teeth on the fifth toe." The shoulder, heart, and lungs have pressure points along the ball of the sole. A point on the heel is connected to the sciatic nerve. And so on.

Dozens of pages display striking pictures showing just how the thumb and fingers apply pressure to reflex points to relieve pain and other symptoms. Reflexology will not cure the causes of such terminal ills as cancer and AIDS, Dougans tells us, but it will ease the pain associated with such ills. Moreover, reflexology, if regularly practiced, will prevent the onslaught of such diseases. You can work on your own feet, as the book explains, but for best results, Dougans insists, you must go to a trained reflexologist. Treatments may take weeks or even months, and some patients respond better than others. Do the officials of Barnes & Noble buy all this balderdash? Of course not! The only thing they believe in is profitology.

I own a rare little book titled *Zone Therapy, or Relieving Pain and Sickness by Nerve Pressure* (1928). The author, Benedict Lust, is a prolific writer who is considered the father of American naturopathy. One of his many books is a profusely illustrated *Universal Naturopathic Encyclopedia*. It has 1,416 pages. Lust ran the American School of Naturopathy and Chiropractic, at 236 East 35th Street, New York City. The school published a periodical, *Nature's Path*, devoted to ways of healing without having to see a physician. In later years Bernarr Macfadden, with his *Physical Culture* magazine and his many books, including a five-volume *Encyclopedia of Physical Culture*, became the nation's top promoter of naturopathy.

Lust follows Dr. Fitzgerald in using elastic bands and spring clothespins to apply pressure on fingers and toes. To prevent falling hair he recommends "rubbing the fingernails of both hands briskly one against the other in a lateral motion, for three or four minutes at a time, at intervals throughout the day. This stimulates nutrition in all the zones, and brings about a better circulation of the entire body, which naturally is reflected in the circulation of the scalp itself."

Amazingly, reflexology is enjoying a mild revival. In my small hometown of Hendersonville, North Carolina, a woman reflexologist was favorably interviewed recently by our local newspaper. The March 29, 1998, issue of *Parade,* the Sunday newspaper supplement, ran a half-page ad for a video titled *Reflexology: The Timeless Art of Self Healing.* Your $22.96 also gets you a free wallet-size reflexology chart.

"Pinch your toes to relieve your sinuses?" the *Parade* ad asks. "Press your heel to ease sciatica? Massage the sole of your foot to quiet a nervous stomach? . . . According to those who practice reflexology, the soles of your feet are maps of your entire body. They contain thousands of tiny nerves called 'reflexes' which correspond to every organ, gland, and bodily function. By pressing or massaging these specific points . . . you can revitalize and balance your body's energy, promote natural healing, and more. Now renowned Reflexologist Ann Gillanders tells you just how beneficial it can be with this easy-to-follow video!"

A more recent ad in *Parade* (January 3, 1999) manages to combine reflexology, acupressure, and magnet therapy. For $12.90 you can buy a pair of "Therasoles." These are described as "magnetic acupressure shoe insoles." Each Therasole has over five hundred "acupressure nubs" designed to "stimulate the nerve endings in your feet that correlate to all parts of your body." As an extra bonus, each Therasole contains five "strategically placed" magnetic disks. Has *Parade* no shame?

Although reflexology is usually confined to working with pressure on the feet, recent claims have been made for healing by pressure and massage on the hand. In the late 1970s and early 1980s the *National Enquirer* and other tabloids ran full-page ads, even double spreads (December 20, 1977, for example), for a book by Mildred Carter titled *Hand Reflexology: Key to Perfect Health.* The ads were headed in large type: "Now! Let me show you how hand reflexology can bring you instant relief from pains all over the body. . . . Cure specific ailments! Simple method requires no expense . . . no special equipment (just your hands). . . . Can be used by anyone in perfect safety!"

"Yes, Dear friend," the ad begins, "I want to tell you about a method that can stop pain in all parts of the body instantly. . . . No matter what sort of ailment you suffer, merely by pressing or rubbing certain relief centers in your hands, you can absolutely relieve the pain, and in most cases, relieve a cause!" A chart shows the hand's pressure points. Testimonials of miraculous healing, taken from Carter's book, are even more sensational than those at the back of Mary Baker Eddy's *Science and Health.*

A lady of forty-five, deaf since age six, heard normally after twenty minutes of hand rubbing. A uterine fibroid vanished instantly. Bronchial asthma and a painful stomach ulcer were cured in five minutes. Carter says she brought her senile mother, aged eighty-six, back to normal by rubbing her hand. A gallbladder pain instantly disappeared. Carter claims instant healing of hangovers, hemorrhoidal pain, toothaches, impaired vision, prostate trouble, and multiple sclerosis. A cataract's growth was halted. The list of such cures goes on and on. A photograph in the ad shows a man stopping his falling hair by rubbing his fingernails together in precisely the manner prescribed by Lust.

On October 26, 1980, the prestigious *New York Times Book Review* ran a full-page ad for a new variant of acupressure called "myotherapy." This unusual therapy was discovered and named by Bonnie Prudden, head of the Myotherapy Institute, in Stockbridge, Massachusetts. She is identified as a "world famous physical fitness expert" who ran a television exercise program. The ad says she discovered myotherapy in 1976 while working with a Dr. Desmond Tivy in the field of "trigger point injection therapy," whatever that is.

Bonnie's ad has a long list of chronic pains that are erased by pressure on what she calls the body's "trigger points." Unlike acupuncture, acupressure, zone therapy, and reflexology, myotherapy holds that each person has a different "map" of trigger points. Her book *Pain Erasure: The Bonnie Prudden Way* (New York: M. Evans & Co.) explains how to locate your trigger points by applying pressure on your body at various spots until you feel a sharp pain. Apparently myotherapy never caught on.

A cartoon I clipped from *Argosy* (June 1974) shows a doctor speaking across his desk to a patient. "Orthodox medicine has no known cure for your condition," he is saying. "Fortunately for you, I'm a quack."

Addendum

Bruce Barton, an American journalist best known for his popular life of Jesus titled *The Man Nobody Knows,* had this to say about zone therapy in *Everybody's Magazine,* which he edited. I quote from Lust's book:

> For almost a year Dr. Bowers has been urging me to publish this article on Dr. Fitzgerald's remarkable system of healing known as zone therapy. Frankly, I could not believe what was claimed for zone therapy, nor did I think that we could get magazine readers to believe it. Finally, a few months ago, I went to Hartford unannounced, and spent a day in Dr. Fitzgerald's offices. I saw patients who had been cured of goiter; I saw throat and ear troubles immediately relieved by zone therapy; I saw nasal operations performed without any anesthetic whatever; and—in a dentist's office—teeth extracted without any anesthetic except the analgesic influence of zone therapy. Afterward I wrote to about fifty practising physicians in various parts of the country who have heard of zone therapy and are using it for the relief of all kinds of cases, even to allay the pains of childbirth. Their letters are on file in my office.
>
> This first article will be followed by a number of others in which Dr. Bowers will explain the application of zone therapy to the various common ailments. I anticipate criticism regarding these articles from two sources: first, from a small percentage of physicians; second, from people who will attempt to use zone therapy without success. We have considered this criticism in advance and are prepared to disregard it. If the articles serve to reduce the sufferings of people in dentists' chairs even ten per cent, if they will help in even the slightest way to relieve the common pains of every-day life, they will be amply justified.
>
> We do not know the full explanation of zone therapy; but we do know that a great many people have been helped by it, and that nobody can possibly be harmed.

Here are some other books on reflexology published in the United States since 1980:

Bayly, Doreen. *Reflexology Today: The Stimulation of the Body's Health Through Foot Massage,* 1984.

Byers, Dwight. *Better Health with Foot Reflexology,* 1983.

Carter, Mildred. *Helping Yourself with Foot Reflexology,* 1986; *Hand Reflexology: Key to Perfect Health,* 1986; *Body Reflexology: Healing at Your Fingertips,* 1983, revised 1994.

Gillanders, Ann. *The Joy of Reflexology,* 1995.

Hall, Nicola. *Reflexology: Foot and Hand Massage for Relaxation and Treating,* 1991; *Reflexology: A Patient's Guide,* 1988.

Issel, Christine. *Reflexology: Art, Science, and History,* 1990.

Kunz, Kevin and Barbara. *The Complete Guide to Foot Reflexology,* 1982, revised 1990; *Hand and Foot Reflexology,* 1992.

Norman, Laura, and Cowan, Thomas. *Feet First: A Guide to Foot Reflexology,* 1988.

Rick, Stephanie. *Reflexology Workout: Hand and Foot Massage for Super Health and Rejuvenation,* 1980.

Special mention should be made of *New Choices in Natural Healing,* edited by Bill Gottlieb, editor in chief of *Prevention* magazine's series of health books. It was first published by Rodale in 1995 and reprinted by Bantam in 1997. This deplorable book sings the praises of reflexology, with maps of the reflex points on both feet and hands and a list of sixty-eight different ailments said to be helped by reflexology.

CHAPTER 9

Urine Therapy

The preceding chapter was about reflexology, the technique of eliminating pain and other symptoms of illness by applying pressure to various spots on the foot. This one concerns an equally crazy therapy that also is currently bamboozling gullible persons captivated by alternative medicines.

From ancient times, folk superstitions have involved the fancied healing properties of three bodily secretions: saliva, excrement, and urine. Saliva has been by far the most popular. You may recall from gospel accounts how Jesus restored sight to a blind man (John 9, Mark 8) by first applying his saliva to the man's eyes. In John's account, Jesus mixes saliva with clay, which he then puts on the man's eyes. In Mark's account, Jesus puts his saliva directly on the eyes.

Mark 7 tells how Jesus healed a deaf man with a speech impediment by first thrusting his fingers into the man's ears, then placing saliva on the man's tongue. Similar accounts of healing with saliva abound in the myths of Hinduism and Buddhism and in medieval legends about Christian saints and their miracles.

The *Encyclopaedia of Religion and Ethics,* edited by James Hastings, has five pages in Volume 11 on worldwide superstitions about saliva. The Roman scholar Pliny the Elder, in his thirty-seven–volume *Historia Naturalis (Natural History),* A.D. 77, describes many curative powers of saliva that were widely believed in his day. Here are a few examples: If you are bitten by a mad dog, drinking the dog's saliva will forestall hydrophobia; horse saliva will cure ear ailments, chafing caused by riding horses, and diseases of the uterus; rubbing spit behind the ear gives peace of mind. Scores of other saliva superstitions are attributed to various cultures. The medicinal properties of the saliva of India's holy men are also cited. For a good account of medieval miracles attributed to spittle, see E. Cobham Brewer's *A Dictionary of Miracles.*

Saliva is still regarded as a powerful healing agent in ayurvedic medicine, India's ancient folk medicine, which is strongly promoted by Maharashi Mahesh Yogi, the guru of Transcendental Meditation, as well as in many books by Deepak Chopra. According to Kurt Butler, in his *Consumer's Guide to Alternative Medicine* (Prometheus Books, 1992), Chopra claims that you can prevent and even reverse cataracts by brushing your teeth, scraping your tongue, spitting into a cup of water, then washing your eyes repeatedly with the mixture. A cataract is the irreversible clouding of an eye lens. You can no more reverse the clouding than you can unfry an egg.

The reputed healing properties of saliva, excrement, and urine are detailed ad nauseam in Lynn Thorndike's monumental *History of Magic and Experimental Science.* Here I shall be concerned only with urine, either swallowed, injected, or applied externally to the skin. Following are a few highlights from Thorndike.

Pliny the Elder, in *Natural History,* extols the healing powers of urine obtained from a virgin boy. Arnald of Villanova, a thirteenth-century Spanish astrologer, alchemist, and physician, claimed that warts would vanish if a dog's urine was applied to them. He also said vision would be greatly improved by washing one's eyes every morning with one's urine. Thorndike further cites an ancient Arabic treatise discussing the healing power of a white elephant's urine.

Urine was a popular folk medicine throughout the seventeenth century. Thorndike gives many references. Emmanuel König, of Basel, in his book *The Animal Kingdom* (1683), recommended drinking one's urine to heal heartburn, depression, gout, toothaches, colic, jaundice, and high fevers. Daniel Bockher, a German physician, in 1622 published a popular work titled *Medicus Microcosmos.* It praises the healing properties of urine, excrement, lice, sperm, tapeworms, and ear wax.

Pierre Fauchard, a Parisian dentist, is considered the founder of modern dentistry. In 1728 he published *Le Chirurgien Dentiste (The Surgeon Dentist),* a classic text translated into English in the 1940s. In this work he pokes fun at several ridiculous remedies for toothache, then proceeds to describe a novel remedy of his own. I quote from James J. Walsh's entertaining *Cures: The Story of the Cures that Fail* (1923):

> I have brought a great deal of relief to a number of patients who had nearly all their teeth carious and who as a consequence were often tormented by pains and aches . . . by means of the following remedy. It consists of rinsing out the mouth every morning and also evening . . . with some spoonfuls of their own urine, just after it has been passed. . . . it is true that it is not very agreeable, except inasmuch as it brings distinct relief.

Fouchard goes on to say that some of his patients who used this remedy told him that the urine also provided relief from other health problems. Indeed, "experience has shown that urine of healthy persons is very good for relieving the pains of gout and getting rid of obstructions of various kinds throughout the body."

Robert Boyle (1627–91) was a great British scientist best known to physicists today for "Boyle's law," which states that the volume of a gas varies inversely with its pressure. Thorndike quotes the following passages from Boyle's *Works:*

The medical virtues of man's urine, both inwardly given, and outwardly applied, would require rather a whole book, than a part of an essay, to enumerate and insist on. . . . I shall now only add, that I knew an ancient gentlewoman, who being almost hopeless to recover of divers chronical distempers . . . was at length advised, instead of more costly physick, to make her morning draughts of her own water; by the use of which she strangely recovered, and is, for aught I know, still well. And the same remedy is not disdained by a person of great quality and beauty, that you know; and that too after she had travelled as far as the *Spaw* for her health's sake.

Urine therapy is most widely used today in Hindu folk medicine. William Jarvis, writing in the newsletter of NCAHF, the National Council Against Health Fraud (March/April 1995), reported that India's former prime minister (1977–79) Morarji Desai on his ninety-ninth birthday, in February 1995, attributed his longevity to a constant drinking of his own urine. In 1978, Dan Rather, on CBS's *60 Minutes,* interviewed Desai, who spoke at length about the great value of drinking urine. *Newsweek* reported (August 21, 1995) that Mohandas Gandhi was a urine drinker, but this was later denied by India's Gandhi Institute.

In the July/August 1991 issue of the *NCAHF Newsletter,* Jarvis cites numerous urine-based remedies that are promoted by ayurvedic medicine. Alcoholism, anorexia, nausea, poor digestion, edema, and other ills respond to "goat feces prepared by washing in urine." For constipation, drink a mixture of milk and urine. Epilepsy and other seizures yield to donkey urine. Urine is called the "water of life" in ayurvedic medicine. G. K. Thakker, director of the Water of Life foundation in Bombay, believes that urine drinking can cure every illness from the common cold to cancer.

Helen Kruger, in her excellent book *Other Healers, Other Cures: A Guide to Alternative Medicine* (1974), has this to say about urine therapy:

The Chinese used to drink the urine of a young boy as a curative. In some parts of the South, a baby's face is washed with urine to protect the skin. Elsewhere, it's used as a gargle for sore throat and for acne, cuts, and wounds. (Don't scoff. Doctors sometimes prescribe an ointment containing *urea,* the chief component of urine, for skin problems.) A Brooklyn woman now in her sixties recalls having had to gargle with her own urine when she had diphtheria as a child. The French had a custom of soaking a stocking in urine and wrapping it around the neck to cure a strep throat. And in the Sierra Madre, the Mexican peasants prepare a poultice for broken bones by having a child urinate into a bowl of powdered charred corn. The mixture is made into a paste and applied to the skin. People have put urine into the eyes to "cure" cataracts, a habit that gives my eye doctor the shudders. It could cause the cataract to grow denser, he says. And only recently, I heard about a new reducing treatment given by some doctor in Florida. The urine of a pregnant woman is injected into the obese patient to "break down the fat cells." (At the same time, however, the patient is put on a near-starvation diet.) Finally, I was told by a healer in Texas (who heard it from "a reliable source" who *knows*) that certain Hollywood beauties of mature years keep their youthful appearance by, as she put it, "drinking their own output."

Among a spate of recent books on urine therapy, one of the most impressive is Martha M. Christy's oversize paperback, *Your Own Perfect Medicine: The Incredible Proven Natural Miracle Cure That Medical Science Has Never Revealed!* This book was published in 1994 by Wishland, Inc. My copy is a 1998 fifth printing! You can obtain it from Wishland, P.O. Box 13927, Scottsdale, AZ 85267 (toll-free 1-800-554-CURE) for $19.95 plus $5 for shipping and handling. The company also sells Christy's *Healing Yourself with Homeopathy* and *Scientific Validation of Urine Therapy,* and her audiotape titled *The Scientific Validation of Urine Therapy.*

Among blurbs on the back cover of *Your Own Perfect Medicine* are the following remarks by David G. Williams, M.D., of the *Alternative Health Newsletter.* "If you buy only one health book this year, this is the book you should get. It outlines a therapy that can be used by anyone, anywhere, for practically any complaint known." Another comment, "This agent was found to be definitely life saving," is attributed to Dr. Manucher J. Javid, identified as a neurosurgeon at the University of Wisconsin's Medical School, in Madison.

I wrote to Dr. Javid, now retired, to ask what he thought of Christy's book. As I suspected, he was appalled by how she had quoted him out of context. His research, he pointed out in a letter, was on the use of urea to reduce intercranial pressure. Urea, he explained, is not obtained from urine. Calcium cyanamide is dissolved in water, then heated under high pressure to produce urea and calcium hydroxide. Dr. Javid added that he has contacted an attorney to consider legal action for using his name without his knowledge or permission. "I can unequivocally state," he writes, "that the use of urea in medicine should not be extended to endorse any claim for drinking or injecting human urine."

In the first chapter of her book, Christy says that between the ages of eighteen and thirty she was diagnosed with pelvic inflammatory disease, ulcerative colitis, ileitis, chronic fatigue syndrome, Hashimoto's disease, and mononucleosis. In addition to these ills, "I had severe chronic kidney infections, two miscarriages, chronic cystitis, severe candida and external yeast infections along with marked adrenal insufficiency and serious chronic ear and sinus infections. . . . And even though I ate almost nothing because of my extreme food allergies, I actually kept gaining weight. . . . I had become nothing more than a walking encyclopedia of disease."

I was reminded of Henny Youngman's joke about the man who, after his doctor examined him, asked, "How do I stand?" "That," said the doctor, "is what puzzles me."

Christy first tried to heal herself using various methods, including a nutrition regime promoted by Adelle Davis, megavitamin therapy, acupuncture, chiropractic, "and every herbal preparation and drug-free natural health therapy" she could find. She says her health started to improve, but after a difficult childbirth all her old ailments returned. Doctors told her the symptoms were psychosomatic. She says she alienated every doctor in town by insisting otherwise. She came down with endometriosis—bleeding tumors that result when uterine tissue detaches from the uterus and attaches itself elsewhere. After five operations, she refused to have a sixth.

Instead, she checked into an alternative cancer clinic in Mexico. The treatments there failed as well. "My usual herbs and homeopathic remedies, although they gave temporary relief, seemed almost useless." For a year she remained in bed. Even though she had health insurance, she and her husband spent more than $100,000 of their own money on alternative therapies. After another tumor was removed, the endometrial symptoms returned, complicated by a yeast infection and menopausal symptoms.

Desperately ill and severely depressed, even contemplating suicide, Christy came across a book that recommended drinking one's urine. She gave it a try. The result was "almost instantaneous relief" from her symptoms. Her hair, which had fallen out, grew back again "thick and lustrous." She gained weight and her energy returned. She says she now swims, hikes, and rides horseback. "Much to my own and my family's amazement, I am back to work and after thirty years of almost nonstop illness, I have a rich, full life again—and all because of an unbelievably simple and effective natural medicine."

The rest of Christy's book reveals in vast detail how drinking one's urine will cure cancer, multiple sclerosis, malaria, arthritis, AIDS, gonorrhea, jaundice, ringworm, tuberculosis, migraine, hepatitis, whooping cough, hay fever, depression, prostate trouble, diabetes—you name it, urine will cure it. The main component in urine that does all these medical wonders is urea.

Christy correctly calls attention to many drugs now on the market that are based on urea, but she insists that these drugs cause unpleasant side effects that are avoided if you obtain the urea by drinking your own urine. She highly recommends a 1971 book, *The Water of Life,* by John W. Armstrong, available from Home Cure for $11.95 plus $5 for shipping and handling. She says it lists hundreds of ailments, including cancer, cured by urine drinking.

Urine therapy, Christy explains, begins by swallowing only one or two drops a day. Even these few drops, she writes, are very effective. "If you prefer, you can make an extremely diluted form of urine called a homeopathic urine preparation, which gives excellent results and contains no taste nor color." Of course it is tasteless and colorless! All the urine has vanished into the water!

Surgery and mainstream drugs simply don't work, Christy assures her readers. Herbal medicines, homeopathic drugs, and acupuncture do work, but none of these therapies, she says, holds a candle to the powerful curative properties of urine. It is "the most powerful, most individualized natural medicine we could ever hope for." Although pure urea has proved to be a wonderful drug, "I want to stress that it cannot and should not be used to replace or supersede natural urine as a healing agent." Why? Because "whole urine contains hundreds of known and unknown medically important elements . . . not found in urea alone."

After a period of swallowing a few drops of urine daily, Christy recommends slowly increasing the number of drops until you finally are drinking several ounces each day. In addition to the value of urine taken internally, Christy also sings its praises when applied externally to the skin and used in foot baths, fully-body baths, ear, nose, and eye drops, and enemas. In cases of extreme illness, Christy writes, injection of urine can be called for. She recommends a Dr. William Hitt who runs two urine therapy clinics in Mexico where "he has administered hundreds of thousands of injections to severely ill patients with remarkable success."

Homeopathic urine, Christy says, is extremely potent. The recipe is simple. To one-sixth of an ounce of distilled water, in a sterile bottle, add one drop of urine. Cap the bottle and vigorously shake it fifty times. A drop of this mix is then added to another sixth-ounce of distilled water. Again, shake fifty times. The third dilution is a drop of the mix into a sixth-ounce of 80 to 90 proof vodka. The vodka, she says, "acts as a preservative." Three drops are placed hourly on the tongue until the symptoms of an illness disappear. Christy is also enthusiastic about combining urine drinking with standard homeopathic drugs.

I know of no books defending urine therapy that are not written by persons who also defend other forms of alternative medicine. Beatrice Barnett, coauthor with Margie Adleman of an addle-pated book titled *The Miracles of Urine Therapy* (1987), is a chiropractor and naturopath. According to Jack Raso, in *Alternative Healthcare: A Comprehensive Guide* (Prometheus 1994), Barnett and Adleman list the following side effects of urine drinking: nausea, vomiting, migraine, boils, pimples, rashes, palpitations, diarrhea, uneasiness, and fever. But they add that these are "normal" symptoms that one should not worry about!

I would have thought that Andrew Weil, the latest guru to be much admired by those who are down on mainstream medicine, might endorse urine drinking. I was pleased to learn that he does not. In his recent book *Ask Dr. Weil* (1998) he says that although urine may have some value when applied to the skin, drinking it has no value whatever. He cites Christy's book *Your Own Perfect Medicine* as an example of a work not to be trusted.

I do not know whether drinking urine is harmless or not, and would welcome hearing from any knowledgeable physician on this point. You can be sure, however, that there are no toxic effects from drinking homeopathic urine. The reason is simple: You are drinking nothing but distilled water.

I also do not know if Christy's book and others like it are cruel hoaxes written and published to make money, or whether the authors believe what they claim. In either case, I shudder at the thought of readers who are seriously ill, and who may be so persuaded that drinking urine will cure whatever ails them that they will not seek medical help that could save their lives.

Addendum

Hugh Trotti, Jr., sent me the following passage from David Rankin's *Celts and the Classical World* (1998):

> A minor, but peculiar, anthropological fact about them, which is very alien to Greek and Roman practice, was probably recorded by Posidonius, though the poet Catullus seems to have known of it: the Celtiberians used stale urine to clean their teeth.

A Reuter news release in 1996 reported on the first world conference on urine therapy, held in Panaji, India. Six hundred devotees gathered to extol the medicinal virtues of human urine when swallowed, gargled, and used as an aftershave. A long list of ailments, including cancer and kidney failure, were cited as cured by drinking urine.

One of the questions asked of Joe and Teresa Graedon, who write a weekly newspaper column titled "The People's Pharmacy," was what they thought of Christy's book *Your Own Perfect Medicine.* Here is how the Graedons replied on November 11, 1998;

> Drinking urine may seem weird to most Americans, but it is an established practice in traditional Indian and Chinese medicine. Some Europeans advocated its use in the 18th and 19th centuries. Healthy urine is sterile and has been used for wound healing on the battlefield.
>
> Proponents of drinking urine claim it bolsters the immune system, with benefits for a variety of ailments. We can't say for sure whether urine therapy works.

I found this answer, in spite of its limp disclaimer at the end, to be reprehensible, and typical of the Graedons' frequent kind words for dubious alternative remedies.

John Armstrong's *Water of Life,* which rivals Christy's book in its extravagant praise of urine therapy, has gone through many editions and is still in print. Copies of its pages had been sent to me in 1987 by Dr. Nick Beard, a London physician—pages that prompted me to write my account.

An entire chapter of Armstrong's book is devoted to miraculous cures of gangrene. Another chapter tells how urine has cured cancers. A naturopath with no medical training, Armstrong nevertheless claims in his introduction to have cured thousands of patients, including himself, of every ailment known simply by drinking urine.

"It is true," he writes, "that at one time I had resolved not to write my book until I had the chance of curing even leprosy; but as I am unlikely to come across a case of this dread disease . . . I have decided to give the details of my experiences to the public without further delay."

Arthur Legate, M.D., responded to my question about whether drinking one's urine is harmful. Uninfected urine, he wrote, is probably harmless, but drugs and various poisons are eliminated by urine. Drinking such infected urine could spread an infection and do more harm than good even if one is dying of thirst in a lifeboat.

Part V

Psychology

CHAPTER 10

Freud's Flawed Theory of Dreams

I have had a most rare vision. I have had a dream, past the wit of
man to say what dream it was: man is but an ass, if he go about
to expound this dream.

—Nick Bottom, a weaver, in Shakespeare's
A Midsummer Night's Dream, *Act 4, Scene 1*

For several decades, Sigmund Freud's reputation as a scientist
has been steadily withering. So much so that *Time*, in 1993, put Freud's
face on its cover, his head depicted as crumbling, and asked: "Is Freud
Dead?" Paul Gray's answer in his feature article was "Yes." Psychiatrists,
philosophers, and critics now regard the "Vienna quack" (as writer Vladimir
Nabokov called him) as a man of great literary talents, but essentially a pseu-
doscientist without the foggiest notion of how to confirm his conjectures.

Nowhere is this paradigm shift more evident than with respect to Freud's
dream theory. Freud himself considered this his finest achievement. In the
preface to the third edition of *The Interpretation of Dreams,* he wrote: "It
contains, even according to my present-day judgment, the most valuable

of all the discoveries it has been my good fortune to make. Insight such as this falls to one's lot but once in a lifetime."

In one of his lectures Freud called his dream theory "the royal road to a knowledge of the unconscious; it is the secret foundation of psychoanalysis." Shortly after his book on dreams was published, he wrote to his close friend Wilhelm Fliess, a bumbling ear, nose, and throat doctor and numerologist from Berlin, that maybe someday a marble tablet would be placed on his (Freud's) house to commemorate where he made his monumental dream discovery. (See *The Complete Letters of Sigmund Freud to Wilhelm Fliess, 1887–1904,* Harvard University Press, 1985.)

Much earlier efforts had been made to unravel dreams. To the ancients, as to today's parapsychologists, dreams were often interpreted as precognitions of future events or clairvoyant visions of current, faraway events. Michel Montaigne, essayist, humanist, and skeptic, in one of his essays wrote: "I believe it to be true that dreams are the true interpreters of our inclinations; but there is art required to sort and understand them."

Before 1900 the prevailing opinion among psychologists was that dreams are mostly random images as nonsensical as Alice's dreams of Wonderland. In Freud's words, they were thought to resemble the sounds of "unskilled fingers wandering over the keys of a piano."

However, Freud also believed that beneath what he called the *manifest* content of a dream—its seemingly absurd, disconnected images—lay a *latent* content that was a cleverly disguised expression of unconscious wishes. "We do literally deny," Freud wrote in his *General Introduction to Psychoanalysis,* "that anything in the dream is a matter of chance or of indifference."

Because most unconscious desires are shocking to the conscious mind, our brain contains something Freud called the "censor." To prevent us from awakening in horror or disgust over an explicit revelation of an unconscious wish, this "severe little manikin" distorts the dream by transforming our secret desires into harmless symbols that will not disturb our slumber. Occasionally, when the censor fails to do its job, the result may be an anxiety dream or nightmare so disturbing that it wakes us.

Freud of course could not deny that dream symbols reflect recent events we have experienced, or even conditions occurring while we sleep, such as unusual heat or cold, loud sounds, strong odors, a stomachache, arthritic pains, and so on. If our bladder is too full, we may dream of urinating—the censor's trick to keep us asleep. If hungry, we may dream of eating; if thirsty, of drinking. In such cases the manifest and latent contents of a dream become the same.

The psychoanalyst's task, helped by free-association tests and dialogue, is to uncover the secret content of a patient's dreams—an indispensable aid in determining the childhood sources of his or her neuroses.

The best introduction to Freud's dream symbolism is Lecture 10 of his *General Introduction to Psychoanalysis.* It must be read to be appreciated.

Male sex symbols are any items that resemble a penis: sticks, umbrellas, poles, trees, knives, daggers, lances, sabers, guns, pistols, mushrooms, keys, pencils, pens, hammers, screwdrivers. Freud doesn't mention bananas, hot dogs, or cigars, but their phallic symbolism is obvious. (Freud is alleged to have once said that in some dreams a cigar may be just a cigar.) Fish and reptiles, especially snakes, are male symbols. So are swans, with their long necks. Neckties that "hang down" and feathers that "stand up" are other male symbols. In *The Interpretation of Dreams,* Freud reports a patient's dream about a hat with a slanted feather. It symbolized the male dreamer's impotence.

Hats and coats can be either male or female symbols. This may be "difficult to divine," Freud writes, "but their symbolism is quite unquestionable." Hats are male symbols because the head goes into them, and coats, because arms go into sleeves. The hats and sleeves also serve as female symbols.

Objects from which water emerges signify male ejaculation: faucets, watering cans, springs, fountains. Anything that flies through the air is symbolic of erection: balloons, airplanes, zeppelins. Common dreams of flying like Peter Pan are dreams of erection. This, Freud tells us, has been proved true "beyond doubt." How is it, then, that women also dream of flying? Freud gives two reasons: They have "penis envy"—a desire to be a man "whether conscious of it or not"—and a woman's clitoris also becomes erect when sexually stimulated.

Female symbols are hollow or they enclose: pits, caves, jars, bottles, boxes, chests, cupboards, shoes (including horseshoes!), slippers, drawers, pockets, jewel cases, ships, stoves, houses, rooms, churches, doors, gates, chimneys, keyholes.

More mysterious female symbols include wood, paper, tables, books, and flowers. If a man dreams of taking flowers from a woman it symbolizes his wish to deflower her. Such puns often play symbolic roles in dreams. A woman dreams of violets. In *The Interpretation of Dreams,* Freud associates this with the French *viol,* meaning rape. Carnations are linked to "carnal." Lilies of the valley are double female symbols because they combine blossoms with valleys.

Snails and mussels, Freud tells us, are "unmistakable female symbols." So are peaches, apples, melons—any kind of fruit that resembles a breast. Female pubic hair is represented in dreams by woods and thickets. When women dream of landscapes, says Freud, the scene swarms with sex symbols: rocks and trees for men, woods for women, and water for both sexes.

Buildings can be either male or female symbols. If outside walls are smooth, the building represents a man with his flat chest. "When there are protuberances such as ledges and balconies which can be caught hold of," Freud writes, the building signifies a woman with projecting breasts.

Both Freud and Carl Jung were fascinated by number symbolism, especially Jung, who carried numerology to preposterous heights. To give only one example, the number 3 signifies for Freud the male genitalia because the figure "3" combines a penis with two testicles. In dreams this is often disguised as a three-leaf clover or the French fleur-de-lis.

A desire to masturbate is represented in dreams by any kind of play, especially piano playing. (Freud would have had a field day with Adelaide Proctor's popular poem and song "The Lost Chord.") Dreams of pulling off branches or having one's teeth yanked symbolize castration as punishment for masturbation.

What about the sex act? For Freud, dreams disguise this as dancing, riding, climbing, or experiencing any kind of violence, such as being run over. Climbing stairways, ladders, or mountains Freud considers "indubitably symbolic of sexual intercourse." He calls attention to the rhythmic aspect of climbing and to its escalating excitement that puts one out of breath.

Throughout his books Freud provides hundreds of examples of dream analysis, often of his own dreams, although he seldom reveals himself as the dreamer. At the time he invented his dream theory, he was a heavy user of cocaine. The drug suppresses dreaming for a time, but there is always a rebound when dreams become more frequent and unusually vivid. Freud carefully wrote down these dreams and did his best to interpret them.

Here is a typical example of how Freud interprets a patient's dream in Lecture 12 of his *General Introduction to Psychoanalysis*. A woman dreams that her head bleeds after banging it against a chandelier. The chandelier is a penis symbol. Her head represents the lower part of her body because as a child her mother once told her that if she didn't behave she would become as bald as her buttocks. "The real subject of the dream then is a bleeding at the lower end of the body, caused by contact with the penis."

Another dream from the same lecture: A woman dreams of seeing a hole in the ground where a tree has been uprooted. Freud has "no doubt" that this dream expresses her infantile belief that she once had a penis, but it had been removed.

Freud theorized that dreams are often what he called "counterwish dreams." These are unpleasant dreams that express fears rather than wishes. For example, a lawyer dreams of losing a case he wants to win, or a woman dreams of being unable to host a banquet she wants to host. In *The Interpretation of Dreams,* Freud recalls the "cleverest" of all his dreamers. This woman strongly wanted to avoid a vacation with her mother-in-law, yet she dreamed of just such a vacation.

One might have expected Freud simply to admit that dreams can reflect fears as well as desires, but no—he struggled all his life to find ways of seeing unpleasant dreams as secret wish fulfillments. He was aware that such counterwish dreams presented serious obstacles to his theory. Here is how he interpreted the unpleasant dream about the vacation with a mother-in-law. The dreamer was in a stage of intense resistance to her analysis. Eager to prove Freud wrong, her unconscious concocted a dream that contradicted his theory! Indeed, Freud found such dreams common among rebellious patients who knew something about psychoanalysis. Of course, there also are "obliging dreams" by knowledgeable patients who want to *please* their analyst.

How about soldiers who in dreams relive horrible traumas they would prefer to forget? These, too, Freud explained as wishes. In such dreams of terror the sleeper is a masochist who wants to continue suffering! As Freud wrote, "Even dreams with a painful content are seen to be wish fulfillments."

Although Freud believed that "an overwhelming majority of symbols in dreams are sexual," he recognized hundreds of nonsexual symbols. Parents are represented by kings, queens, and other authority figures. Brothers and sisters are symbolized by little animals and vermin. Birth is "almost invariably" represented by water, a symbol of the amniotic fluid. Long journeys signify dying.

In the 1920s and 1930s, when Freud was most fashionable in the United States, his devotees had great fun searching for sex symbols in their dreams, and in art and literature. Today's psychiatrists, aside from elderly analysts who still view Freud's writings as sacred, regard Freud's theory of dreams not as his greatest achievement but as his greatest failure. The symbolism is so flexible that a clever analyst, on the basis of data gained from couch dialogue and free-association tests, can interpret any dream to fit any conjecture. A good example of such elasticity was Freud's belief that any dream can stand for its direct opposite "just as easily as for itself." A male symbol can denote a female, and vice versa! Yet Freud, in his vast hubris, was so blind to the absurdities of his dream theory that he expressed amazement that his theory met such "strenuous opposition amongst educated persons."

Sir Peter Medawar, the distinguished British biologist, writer, and Nobel Prize winner, reviewing a book on psychiatry in *The New York Review of Books* (January 23, 1975), concluded:

Psychoanalysts will continue to perpetrate the most ghastly blunders just so long as they persevere in their impudent and intellectually disabling belief that they enjoy a "privileged access to the truth." The opinion is gaining ground that doctrinaire psychoanalytic theory is the most stupendous intellectual confidence trick of the twentieth century: and a terminal product as well—something akin to a dinosaur or a zeppelin in the history of ideas, a vast structure of radically unsound design with no posterity.

The American writer Tom Wolfe, in *In Our Time,* put it this way:

Freudianism was finally buried by the academic establishment in the 1970s, ending its forty-year reign in the United States. By 1979 Freudian psychology was treated only as an interesting historical note. The fashionable new frontier was the clinical study of the central nervous system, an attempt to map precisely how the panel is wired for fear, lust, hunger, boredom, or any other neural or mental event. Long overshadowed by psychoanalysis, brain physiology came into its own with the development of such equipment as the stereotactic needle implant. Today the new savants probe and probe and slice and slice and project their slides and re-

Time, November 20, 1993 (Time/Life Syndication)

gard Freud's mental constructs, his "libidos," "Oedipal complexes," and the rest, as quaint quackeries of yore, along the lines of Mesmer's "animal magnetism" and "*baquet* processes." The central concept of Freud's pathology, the "neurosis," is now regarded as a laughable historicism on the order of "melancholia" or "phlegmatism." Freud himself is regarded as an unusually humorless quack.

Addendum

My attack on Freud brought a raft of angry letters from dedicated Freudians. One reader assured me that Freudianism is "alive and well." True, but alive and well only among a dwindling remnant of Freud acolytes, not among the majority of today's psychiatrists or intellectuals. He also wondered if I would admit that the unconscious is a useful concept. His implication was that Freud was the discoverer of the unconscious. That this is not true my correspondent would realize if he would bother to read the section on psychosomatic ills in William James's *Principles of Psychology*. And the concept long predates James.

To repeat what is often said: Where Freud was sound he was not original, and where original he was mistaken. Compare with Darwin. Where Darwin was sound he was original. Where he was not original, as in his defense of Lamarkism, he was mistaken.

Another correspondent insisted that Freud was a man who cared deeply and honestly for his patients, and who sought to live up to the high moral standards of a doctor of medicine. Baloney. See my chapter on "Freud, Fliess, and Emma's Nose," in *The New Age* (1988). It tells of Freud's contemptible defense of his friend Fliess's bungled operation on the nose of one of Freud's long-suffering patients.

No reader identified the source of Freud's alleged remark when a student asked if there was any symbolic significance to a cigar Freud was then smoking. Until I learn otherwise, I will consider this an unfounded anecdote.

In 1995 the Library of Congress postponed its planned exhibit on Freud after receiving strong protests from Gloria Steinem, Oliver Sacks, Freud's granddaughter Sophie, and thirty-nine other scholars who signed the petition. Peter Swales led the protest, which claimed that the exhibit's advisory commission was packed with Freudians and that federal funds were being used to stage a public relations campaign.

Time, reporting the flap in its December 18, 1995, issue, quotes science writer Frank Sulloway: "[Freud's] model of the mind and notion of dreaming are in total conflict with modern science. His major edifice is built on quicksand."

As for Freud's ethics, here is what *Time* had to say about it:

> A booming revisionist literature paints Freud as both a bad therapist (he admits in letters that he napped or wrote correspondence during sessions) and a manipulator who either ignored his patients' stories or concocted repressed memories to jibe with his theories. The noted patient Dora, for example, complained she was molested by a family friend. He diagnosed her as repressing memories of childhood masturbation and desire for her molester. He counseled another patient to leave his wife so that Freud might cull donations from the man's wealthy mistress.
>
> Such license doesn't play anymore. Melvin Sabshin, medical director of the American Psychiatric Association, says university psych departments have moved away from Freud toward more empirical fields, especially biology and pharmacology. Meanwhile, insurers and patients have embraced the concrete, if problematic, results offered by drugs. Even undergrads "aren't that interested in Freud," admits Roth. "They just think they know he was wrong."
>
> Which leaves the sitcoms. On a recent *Seinfeld,* the characters quipped about "pouch envy." For all his influence, this may someday be how we think of Freud: an old joke that everybody gets, but no one can remember how important it once was.

The exhibit finally opened three years later. Paul Robinson, reviewing it in the *New York Times* (November 22, 1998), noticed that the Freudians have stopped calling Freud a scientist. They now present him as an "imaginative artist" similar to Shakespeare and Dickens.

Freud's *Interpretation of Dreams* was published in 1899, although 1900 appears on the title page by mistake. In 1909, in a letter to Theodore Flournoy, William James commented on Freud's dream theory as follows:

> I hope that Freud and his pupils will push their ideas to their utmost limits, so that we may learn what they are. They can't fail to throw light on human nature; but I confess that he [Freud] made on me personally the impression of a man obsessed with fixed ideas. I can make nothing in my own case with his dream theories, and obviously "symbolism" is a most dangerous method.

A raft of recent books attack Freud as a crank. Here are some of them: Robert Wilcocks, *Maelzel's Chess Player: Sigmund Freud and the Rhetoric of Deceit* (1994); E. M. Thornton, *The Freudian Fallacy* (1984); Frederick Crews, *Unauthorized Freud: Doubters Confront a Legend* (1998); Edward Dolnick, *Madness on the Couch* (1998); Jeffrey Masson, *The Assault on Truth* (1984); and Adolf Grünbaum, *The Failure of Psychoanalysis* (1984) and *Validation of the Clinical Theory of Psychoanalysis* (1993).

A new translation of Freud's *The Interpretation of Dreams* by Joyce Crick (Oxford, 1999) was reviewed at length by anti-Freudian G. William Domhoff in *American Scientist* (Vol. 88, March–April 2000, pp. 175–78). A psychologist at the University of California, Santa Cruz, Domhoff is the author of *The Mystique of Dreams* (1985) and *Finding Meaning in Dreams* (1996). In his informative review he recommends two surveys of the experimental evidence hostile to Freud. Both are by Seymour Fisher and Roger Greenberg: *The Scientific Credibility of Freud's Theories and Therapy* (1977) and *Freud Scientifically Appraised* (1996).

Post-Freudian Dream Theory

Dreams have a kind of hellish ingenuity and energy in the
pursuit of the inappropriate; the most omniscient and cunning
artist never took so much trouble or achieved such success in
finding exactly the word that was right or exactly the action that
was significant, as this midnight lord of misrule can do in
finding exactly the word that is wrong and exactly the action that
is meaningless.

—*G. K. Chesterton, in* The Coloured Lands

The dream theories of Sigmund Freud and Carl Jung were subjective speculations almost totally without empirical support. Not until
1952 was there a major breakthrough in laboratory investigations of dreams.
That was the year Eugene Aserinsky, a graduate student in physiology at
the University of Chicago, accidently discovered REM, the rapid eye movements that accompany deep-sleep dreaming.

Aserinsky had attached electrodes near the eyes of his sleeping ten-year-
old son Armond. He was surprised to see that the EEG (electroen-
cephalograph) machine was tracing wide swings on its graph paper. Further
research by Aserinsky and the late Nathaniel Kleitman, director of the university's sleep research, made the great discovery that periods of REM were

signs of vivid dreaming in contrast to the feeble dreams of NREM (non-REM) sleep. Kleitman died in 1999 at age 104.

REM sleep, it soon became apparent, occurs in intervals throughout the night, usually four to six times, each lasting from ten minutes to an hour. Subjects who believed they dreamed rarely, or not at all, were amazed to find they had strong memories of dreams when they were awakened during a REM period. New facts came to light: Nightmares and sleepwalking occur only during NREM sleep. The belief that a long dream could last only a few seconds proved to be a myth. Types of food eaten during the day have no effect on REM dreams. Recordings played during sleep have no influence on learning, although such spurious claims continue to be made today for audiotapes widely advertised, even in a few popular science magazines.

Intensive research on REM sleep was taken up in scores of laboratories around the world. It was discovered that almost all mammals so far tested have REM sleep periods (including bats, moles, and whales) except, curiously, Australia's spiny anteater. Reptiles lack REM sleep, but birds seem to have intervals of REM that last a few seconds while their heads are under their wings. Dogs and cats clearly have REM dreams. You can lift a dreaming cat's eyelids and see the eyeballs dart back and forth.

REM dreaming surely serves some useful function, otherwise why would evolution have invented it? Exactly what that function is remains a riddle. One plausible argument is that during the night, when it is difficult to hunt for food, mammals began to rest their bodies and minds until the sun arose. Some mammals even hibernate through cold winters. This, however, sheds little light on the function of dreams.

The computer revolution, and the view of AI (artificial intelligence) researchers that a brain is nothing more than an organic computer, led inevitably to computer-derived theories of dreaming. One of the earliest papers advocating such a theory was "Dreaming: An Analogy from Computers," in *New Scientist* (Vol. 24, 1964, pp. 577–79). The authors were two British scientists: psychologist Christopher Riche Evans and computer expert Edgar Arthur Newman. In 1993, Evans's posthumous work *Landscapes of the Night: How and Why We Dream* was published. His 1973 book *Cults of Unreason* contains a major attack on Scientology.

The Evans-Newman theory is that the brain, like a computer, gets cluttered with useless information. Just as a computer's memory has to be routinely cleaned of unwanted junk, so too does our brain need periodic scrubbing. Dreams are the process by which the sleeping brain moves information worth preserving into its long-term memory and erases from short-term memory the trivia that otherwise would clog neural pathways. Why remember such things as the color of the socks you wore yesterday, or what you had for lunch, or everything you said during idle conversation?

As electrical impulses zip around the brain to eliminate such garbage, the pulses activate adjacent neurons to call up patterns that are essentially random. Our unconscious brain does its best to put these images into some sort of coherent scenario, but because they are randomly accessed, the dream story exhibits bizarre nonsense and abrupt transitions like the scenes in Lewis Carroll's two *Alice* books. For an hour or two every night we go harmlessly insane!

Freud believed that dreams are symbols expressing in heavily disguised form the repressed wishes of the id (unconscious), most of them sexual and going back to childhood. If these wishes were not disguised, Freud believed, our shocked superego, with its moral imperatives, would wake us up.

Carl Jung discarded what he thought was Freud's overemphasis on repressed sexual desires. In his view dreams reflect "archetypes"—memory traces inherited from our evolutionary past. Dreams of flying and falling, for example, are genetic memories of ancestors swinging through trees and occasionally dropping to the ground.

> *It is a wonderous thing to dream*
> *Of tumbling with a fearful shock*
> *From some tall cliff where eagles scream,*
> *To light upon a feather rock.*

So wrote the British poet Martin Tupper in eight stanzas of doggerel titled "Dreams."

Terror dreams of being pursued reflect times when our ancestors fled from fierce beasts. For Jung, dreams do not so much conceal as they reveal these ancient memories buried in what he called humanity's "collective un-conscious."

Evans and Newman have no use for either Freud or Jung. Dreams, they argue, are essentially nonsense, though of course influenced by hopes and fears, and by night events such as sounds, smells, temperature, drafts, bod-ily distresses, and so on. Our brain filters out accustomed noises, such as rain, the hum of an air conditioner, or a television set left on, but sudden, significant sounds, such as a baby's cries, a thunderclap, or a ringing phone, either wake us or are incorporated into a dream. If we are thirsty we may dream of drinking; if hungry we may dream of eating. If our bladder is full, we may dream of urinating. If our face is sprayed with water, we may dream of taking a shower.

Just as it is difficult to remove unwanted data from a computer while it is working on a problem, we would go mad if the sweeping of junk from our memory occurred while we are awake and busy processing new sen-sory inputs. It is not so much that dreams open up storage space, Evans and Newman maintain, as that they clear pathways to provide simpler, more direct access to significant memories. If unwanted data are not routinely removed from a computer, its speed and efficiency are reduced, and the soft-ware may even crash. Similarly, if we are deprived of REM dreaming, we develop behavior disorders and mental distress until allowed to dream again. Instead of dreams preserving sleep, as Freud believed, it is the other way around. We sleep in order to dream.

In the early 1980s, Francis Crick, a Nobel Prize winner for his role in discovering the helical structure of DNA, and mathematician Graeme Mitchison proposed a dream theory similar in some respects to the Evans-Newman conjecture. Their speculations were first presented in "The Func-tion of Deep Sleep," in *Nature* (Vol. 304, July 14, 1983, pp. 111–14). Instead of the brain becoming clogged with junk memories, its neocortex becomes clogged with accidental neural connections.[1] The brain's billions

---

[1] The neocortex is a highly developed part of the cortex—the outside layer of gray mat-ter—that began its evolution with mammals. It is thought to be the region where memo-ries are finally stored, and where reasoning takes place.

of neurons are interconnected in an inconceivably complex web—the most complicated structure known in the universe. When normal memories are stored, the process tends to strengthen unwanted neural connections. Crick and Mitchison call them "parasitic memories." The purpose of REM sleep is to dampen these accidental synaptic connections and so erase spurious memories. Such a random process naturally fabricates bizzare nonsense scenes.

Babies experience twice as much REM sleep as adults, and even show REM in the womb—facts that, if the babies are actually dreaming, seem to contradict Freud's theory. For Crick and Mitchison, babies dream to keep their brains as free as possible of undesirable neuron connections that otherwise would interfere with the efficient formation of memories. The spiny anteater's lack of REM is explained by its unusually large neocortex. Just as neural nets of computers, if huge, can accommodate spurious neural connections without overloading, so can the anteater's oversize neocortex.

Freudians find it useful to recall and analyze dreams. Crick and Mitchison suggest otherwise. "We dream," they write in *Nature*, "in order to forget." Efforts to recall dreams may actually do harm. "Attempting to remember one's dreams should perhaps not be encouraged because such remembering may help to retain patterns of thought which are better forgotten. These are the very patterns the organism is attempting to damp down." (See Theodore Melnechuk's article on Crick's theory, "The Dream Machine," in *Psychology Today*, November 1983.)

Many other conjectures about dreaming have been proposed in recent years, but far and away the best and most influential of recent books on the topic is *The Dreaming Brain* (1989) by J. Allan Hobson, a psychiatry professor at Harvard Medical School. His commonsense views, like those of most of today's dream researchers, are strongly anti-Freud.

Hobson agrees with the two theories just discussed that dreams have no hidden or "latent content," to use Freud's terminology. They have only a "manifest content." They are what Hobson likes to call, echoing Jung, "transparent." Instead of erasing trivial memories, or damping down unwanted accidental neural connections, the brain is merely using its electrical energy to fire neurons more or less randomly while we sleep. In doing so, its images are naturally influenced by recent events (what psychoanalysts call "day residue"), by old memories, by conditions in the bedroom, by body states, and by strong hopes and fears.

Because dreams do not disguise unconscious wishes, no insights into dreams can be gained by free association tests or by trying to interpret outlandish Freudian symbols. Dreams are just what they seem to be. If you dream of missing a train or a plane it is because in life you have experienced such unhappy events. If you dream of a friendly encounter with a relative or other person it is because you are fond of that person. If you dream of an unfriendly encounter it is because you dislike or fear that person. If you dream of flying it is because you often imagine how pleasant it would be to flit about through the air, perhaps reinforced by memories of diving into water, jumping, skating, sledding, and so on.

Freud is said to have remarked that in dreams a cigar may be nothing more than a cigar. For Hobson, a dream cigar is always a cigar. I once had a lucid dream in which I found myself in a strange room with a smoking cigar on an ashtray. Aware that I was dreaming, I decided to experiment to see if in addition to vivid imagery (I could see intricate patterns on wallpaper), my dream could include smells. I picked up the cigar and held it to my nose. The result was such a strong odor of burning tobacco that it woke me up. My dream cigar was only a cigar.

Hobson recalls a vivid dream in which, during a visit to Boston's Museum of Fine Arts, he saw and heard Mozart play a piano concerto. He noticed that Mozart had gotten fat. A Freudian analyst might conclude that Mozart was a father image and that his being overweight symbolized Hobson's unconscious wish to kill his father so he could have his mother to himself. Hobson notes that the concerto was one he knew well. He often listens to Mozart while driving, and he frequently visits the Boston Museum of Fine Arts. His own belly was starting to bulge. The dream had no latent meaning. As Hobson puts it, "Mozart is Mozart." Other books by Hobson are *Sleep* (1989) and *The Chemistry of Conscious States* (1994).

Although we spend a third of our life knocked out, why this is necessary to our health is still unclear. We know sleep refreshes the body, and that somehow it knits up the raveled sleeve of care, as Shakespeare's Macbeth says. The notion once held that dreams "rest" the neurons has to be discarded because neurons are now known to be as active during sleep as when we are awake. Hobson conjectures that the rest theory can be revived if we assume that dreams relax fatigued neurotransmitters in the brain that actually do damp down their firing markedly during REM sleep. Another Hobson conjecture is that maybe evolution developed dreaming partly as a form of entertainment, since most dreams are amusing and delightful, like reading a fantasy tale or watching fantasy plays and movies.

In all three theories the bizarre nature of dreams is explained by haphazard neuron firing, and by the brain's efforts to connect nonsense scenes into a plausible scenario. (This is not the place to discuss lucid or out-of-body dreams in which one is aware that one is sleeping and has a modicum of free will in controlling episodes. See Susan Blackmore, *Beyond the Body: Investigations of Out-of-the Body Experiences,* London: Heinemann, 1982.)

Now that Freud's dream theory is rapidly evaporating like a bad dream, where is dream theory today? Although much is still being discovered, and many rival theories are being proposed, exactly how and why we dream remains a deep mystery. Surprisingly, today's speculations are not much different from what they were for Plato and Aristotle.

Addendum

When I said that nightmares occur only during NREM sleep I was using the term in its older meaning of a dream so fearful that the dreamer cries out in great panic. Several readers wrote to tell me that such dreams are today called "night terrors." A nightmare is considered merely a vivid bad dream, which of course can occur in REM sleep.

Antony Flew and D. F. Hughes each wrote to point out correctly that evolution filters out harmful mutations, often leaving harmless mutations undisturbed even if they have no survival value.

The Paradox of Sleep: The Story of Dreaming, by French scientist Michel Jouvet, was published by the MIT Press in 1999. "Paradoxical sleep" is Jouvet's term for REM sleep; paradoxical because no one really knows why we dream. Sleep is clearly essential for health, but apparently persons can be deprived of REM dreaming for long periods without any adverse effects. Jouvet sides with those who see dreaming as the brain's way of reprogramming itself and erasing uneeded memories.

In reading a review of this book in *American Scientist* (September/October 1999)—I have not read the book—I learned a surprising fact. A dolphin will drown unless it continually breathes air. If fully asleep, it would not be able to periodically surface for breathing. How does it solve the problem of both sleeping and dreaming? It sleeps with one side of its brain, while the other side stays awake to ensure it obtains the necessary air!

Science News (November 8, 1997) and *Discover* (March 1998) reported the astonishing discovery that the platypus, a primitive mammal that lays eggs like the spiny anteater, spends up to eight hours a day in REM sleep! This is more than six times the amount of human REM sleep. The discovery was surprising because the spiny anteater doesn't dream at all, yet its brain structure and that of the platypus are almost identical. Neuroscientist Jerome Siegel, of the University of California, who together with colleagues in Australia did the platypus research, calls the little animal the "REM sleep champion." How all this bears on the evolution of REM sleep and dreaming is far from clear.

J. Allan Hobson's beautifully illustrated book *Sleep* was published by W. H. Freeman in 1989. The book's focus is sleep and its disorders, but it contains a chapter on dreams that summarizes current competing theories.

This is not the place for a bibliography of the many books on dreaming published in the last few decades, but two recent books are worth citing: David Foulkes, *Children Dreaming and the Devlopment of Consciousness* (1999), and Owen Flanagan, *Dreaming Souls: Sleep, Dreams, and the Evolution of the Conscious Mind* (2000).

The Dream, by Pablo Picasso (Scala/Art Resource, New York)

CHAPTER 12

Jean Houston

New Age Guru

Outside New Age circles, the public knew little about Jean Houston until Bob Woodward, in his 1996 book *The Choice,* devoted ten pages to how Houston and Hillary Rodham Clinton became good friends. As everyone now knows, Hillary Clinton had many sessions with Houston during which, as a mental exercise, the First Lady held imaginary conversations with Eleanor Roosevelt and Mahatma Gandhi. She balked at conversing with Jesus, calling such a dialogue "too personal."

As Houston and Mrs. Clinton later made clear, in no way did the First Lady think she was in touch with the spirits of Mrs. Roosevelt and others. It was no more than what Houston calls one of her "mind games." She describes herself as a philosopher and psychologist who never attended a séance and has not the slightest interest in spiritualism.

This is true. Nevertheless Houston and her husband, Robert E. L. Masters, have an abiding interest in channeling. Famous channelers such as J. Zebra Knight claim to be transmitting messages from actual discarnates or entities in higher worlds, but Houston and Masters see channeling from an entirely different perspective. They are convinced that channelers, while in trance, are in contact with what Carl Jung called the "collective unconscious" of the human race. Deep inside our minds are the "eternal archetypes"—unconscious memories created by our evolutionary history—memories that are the sources of great wisdom.

Houston and Masters began their careers by experimenting with LSD and other hallucinogenic drugs as a way of tapping the collective unconscious. Their first book, *Varieties of Psychedelic Experience* (1966), created a sensation among young people then experimenting with such drugs at the urging of the late Timothy Leary. After LSD was legally banned, Houston and Masters turned to nondrug techniques for exploring what they called the "inner self," especially techniques based on hypnotism and mental imaging.

For three decades Houston and her husband conducted thousands of pseudochanneling sessions with subjects at their Foundation for Mind Research, now in Pomona, New York, located in a house built by actor Burgess Meredith. Subjects are put into a trance state during which they seem to speak with the voices of persons long dead. For example, one woman patient transmitted striking messages that purported to come from an ancient Egyptian goddess called Sekhmet. Masters's 1988 book *The Goddess Sekhmet* (reprinted 1991) is about these sessions.

Neither Masters nor Houston thinks an actual Egyptian goddess spoke through the woman's lips. At the same time, they are overwhelmed by the beauty and wisdom of her messages and the messages that come through other "channelers" when they make contact with the collective unconscious.

Hillary Clinton may not be aware of Houston's belief, shared by Masters, that persons in a trance state can have heightened powers of ESP (telepathy, clairvoyance, and precognition).

These views of Houston and Masters are covered in some fifteen books they have written independently or together. The two most influential are *Mind Games: The Guide to Inner Space* (1972) and *Listening to the Body* (1978). Earlier books authored or coauthored by Masters include *The Cradle of Erotica* (1963), about sexual practices in Africa and Asia, and *Eros and Evil* (1962), on the relation between sexual beliefs and the burning of medieval witches.

A good account of the psychic opinions of Houston and Masters can be found in Jon Klimo's *Channeling: Investigations on Receiving Information from Paranormal Sources.* The book was published in 1987 by the New Age publishing firm of Jeremy P. Tarcher, husband of the ventriloquist Shari Lewis.

Houston told Klimo:

> These [channeled] "entities" as we call them—Seth or Saul or Paul or Jonathan—are essentially "goddings" of the depths of the psyche. . . . They are personae of the self that take on acceptable form so that we can have relationship to them and thus dialogue. . . . [T]he traditional archetypes do not have for many people the power they once held. People are in a kind of free-form archetypal search. And so you get the Seths and the Salems and the myriads of very personal guides that are filling the psyche.

In 1979, Ken Carey, a young Missouri farmer, began channeling an entity named Raphael (perhaps the Bible's archangel) and later Jesus himself. Carey published his channeled gibberish in two preposterous books: *The Starseed Transmissions: An Extraterrestrial Report* (1982) and *Vision* (1985). "As I communed with these spatial intelligences," Carey wrote, "our biogravitational fields seemed to merge, our awareness blend, and my nervous system seemed to become available to them as a channel for communication."

Carey's channeled material swarms with the usual New Age buzzwords—unity, vibrations, wholeness, harmony, love, and so on—without conveying anything significant. Nevertheless, Houston assured Klimo that the

Starseed Transmissions are "perhaps the finest example of 'channeled knowledge' I ever encountered."

Although Houston repeatedly denies that channelers are in contact with individuals who have died, she comes close to saying that they are in touch with minds outside their own brains. Here is how she put it to Klimo:

> I think the universe is filled with intelligence. Some is embodied, some is disembodied, and ultimately probably none of it is disembodied. It may have minus n-dimensional structure or go through a black hole to a negative particle structure. But I think everything has structure or pattern. And I think it is how you define embodiment. If it's embodiment of protein or computer or chemicals, then we have a problem. If you think of embodiment as perhaps of frequency which is pulsed, they may not even be in space, they may be across time.

Houston is fond of what is called the *holographic universe.* Like a hologram, every portion of a holographic universe in some way contains the whole. Klimo reports Houston as saying:

> If the holographic theory has any proof to it, then it suggests that everything is ubiquitous [omnipresent] with everything all the time anyway in this simultaneous-everywhere matrix universe. And that would take care of a great deal of the channels. . . . Certain individuals are able to raise the gates or lower the ice or become diaphanous, be stretched very thin, so that this ubiquitous, simultaneous-everywhere-all-the-time information is available to human beings. . . . Then you've got to project a persona [a source] to contain the ubiquity of the information, otherwise you're not going to take it.

The Foundation for Mind Research was originally headquartered in an apartment in Manhattan on East 86th Street. Its laboratory contained, perhaps still does, biofeedback equipment and a sensory deprivation chamber in which subjects meditated in silent pitch darkness. Another piece of equipment was a strobe light box—you would put your forehead against the box, eyes shut, while the flickering light came through your eyelids to enhance meditation. The most bizarre apparatus was called an ASCID (Altered State of Consciousness Induction Device), often referred to as a "witch's cradle." Subjects sat on it blindfolded while it swung back and forth like a pendulum.

Houston and Masters believe that such devices can make one's brain work faster, compressing an hour's work into just a few minutes. A novelist, for example, might see in a flash how to finish a novel he had been trying for months to complete. In the early days of their collaboration, Houston and Masters conducted experiments in dream telepathy with the parapsychologists then running the dream laboratory at Maimonides Medical Center.

Houston likes to tell about an incident that occurred when her father, Jack, was a professional comedy writer for Bob Hope, George Burns, Henny Youngman, and other comics, and for the ventriloquist Edgar Bergen. I quote again from Klimo:

One time my father and I came into Edgar's room. He didn't know we were watching him. Edgar was talking to Charlie and we thought he was rehearsing, but he was not rehearsing. He was asking Charlie questions: "Charlie, what is the nature of life? Charlie, what is the nature of love?" And this wooden dummy was answering quite unlike the being I knew on the radio. A regular wooden Socrates, he was. It was the same voice but it was a very different content altogether. And Bergen would get fascinated and say, "Well, Charlie, what is the nature of true virtue?" and the dummy would just pour out this stuff: beauty, elegance, brilliant. And then we got embarrassed and coughed. Bergen looked around and turned beet red and said, "Oh, hello, you caught us." And my father said, "What were you doing?" And he said, "Oh, I was talking to Charlie. He's the wisest person I know." And my father said, "But that's your mind; that's *your* voice coming through that wooden creature." And Ed said, "Well, I guess ultimately it is, but I ask Charlie these questions and he answers, and I haven't the faintest idea of what he's going to say and I'm astounded by his brilliance—so much more than I know." To me that was a classical channeling instance where the dummy was used as [an] amanuensis of depth structures of Bergen's mind.

An ad for Houston's 1980 book, *Life Force: The Psycho-Historical Recovery of the Self,* is headed, "For the first time in human history we can become fully human." By using the special exercises in this book, the ad continues, "we can relive the whole of human history in our own lives—healing the traumas of ages past and present as we resolve and heal the crises of our own emergent selves. As a result, what may have been lost of the promise and potential of these historical and personal ages may be found again— and used to restore the balance between our inner and outer space."

Houston's 1982 book, *The Possible Human: A Course in Extending Your Physical, Mental, and Creative Abilities,* was reviewed this way in a psychic magazine called *New Realities* (Vol. 5, No. 2, 1983):

> Houston maps out a route leading from the diminished realm of the human-all-too-human to an expansive sensory-intellectual-mystical sphere of the superhuman, what the great contemporary Hindu mystic Aurobindo termed Gnostic being or what Houston likens to the Yiddish archetype, the Mensch.

Houston is tireless in conducting workshops around the world in which she uses a bewildering variety of techniques designed to raise the "human potential" of her students. In 1989 at a conference at the Oasis Center, in Chicago, her workshop was organized around the concept of Pangaea. This is a name for the massive continent that existed 200 million years ago before it broke up into the present continents. Houston took Pangaea to symbolize the primitive races which, in the words of a brochure for the workshop, "contain DNA-like codings that can lead us back to an understanding of who and what we truly are. Jean and her associates devise exercises, processes and practices to reawaken our senses and our memories of these codings throughout our body/brain/mind systems. We find ourselves quickened and empowered to join as partners in the evolutionary process, and to claim our place as true citizens of Panga[e]a. . . . [O]ur time at Oasis this year will be spent on empowering the archetypal creator of, and dweller in, Paradise."

The brochure continues, saying that workshop participants will study "the wise ones of primal cultures such as, perhaps, the Bushmen of the Kalahari and, also, perhaps, those huge-brained mammals who millions of years ago chose to return to the sea, the whales and dolphins."

In *The Possible Human,* Houston says that when she was a junior at Barnard College she was "president of the college drama society, a member of the student senate, winner of two off-Broadway critics' awards for acting and directing . . . and had just turned down an offer to train for the next Olympics." (She had been a fencing expert in high school.) The deaths of family members and friends plunged her into a profound depression. She says she felt like Job, even calling out to God, "Where are the boils?"

Houston attributes her recovery to classes taught by Jacob Taubes, a Swiss professor of religion at Barnard. After that, she writes, "I was off and running and haven't shut up since." After turning down a seven-year contract with Paramount Pictures, she participated in digs in Egypt and Greece, but soon abandoned archaeology to concentrate on what she calls "the archaeology of the self." Houston claims an ability to read Latin, Greek, and Egyptian hieroglyphs. She writes of friendships with such notables as Teilhard de Chardin, Martin Buber, Aldous Huxley, Reinhold Niebuhr, Paul Tillich, Helen Keller, Joseph Campbell, and Margaret Mead.

There is no doubt about her close mother-daughter-like relationship for many years with the elderly and ill Margaret Mead. Mead was a strong believer in psychic phenomena and the reality of visits to Earth by extraterrestrials in UFOs. She became a director of the Foundation for Mind Research. Houston is also a good friend of Mead's daughter Mary Bateson, who accompanied Houston on many of her sessions with Hillary Clinton.

Now fifty-seven, Houston is still a striking woman who likes to say that the Greek goddess Athena is her "archetypal predecessor." Her speech is in a flowery New Age jargon, so vague and murky that it is often difficult to understand. Here, for example, is a typical passage from *The Possible Human:*

Like sunlight and nutrients of earth and air organizing around the buds of springtime, in creative states the budding intentions are in resonance with the holonomic reality, evoking from the primary order the proper nutrients needed to allow for the emergence of the physical expression of the creative intention, be it a novel, a symphony, a dissertation, a relationship, a business, or a community enterprise.

There was what Houston calls her "messianic period," during which she motorcycled "all over east Texas saving people, and preaching and doing 'miracles' like curing people of stuttering." She says she was once thrown out of New York's High School for the Performing Arts "for trying to convert students to a purer lifestyle." On several occasions she has claimed to have a doctorate from Columbia University. According to *Newsweek* ("Soul Searching," by Kenneth Woodward, July 8, 1996), she refused to make changes in her thesis and the degree was never given. Later, in 1973, she did obtain a doctorate in psychology from a school in Cincinnati called the Union Institute. The school was not accredited until 1985.

Houston taught at Marymount College, a Catholic school in Tarrytown, New York, from 1965 to 1972. Although not on the faculty, she was a visiting scholar at the University of Oklahoma in 1980 and a lecturer at Hunter College in 1961.

Houston has close ties with the Apollo 14 astronaut Edgar Mitchell, founder and head of the Institute of Noetic Sciences, in Sausalito, California. Mitchell and members of his institute buy almost everything on the psi scene. Houston says she assisted Mitchell in readjusting to life on Earth after he returned from space. After the flap with Hillary Clinton, Mitchell was quoted as saying: "I would like to support Jean Houston as one of the more worthy and capable scientists, philosophers, and thinkers of our period."

It is usually impossible to know the core beliefs of a president and his wife until long after they leave the White House. Bill Clinton belongs to a Southern Baptist church. Hillary Clinton attends a Methodist church. This tells us nothing about whether either one of them shares the paranormal convictions of Jean Houston. It would be interesting to know.

Part VI

Social Science

CHAPTER 13

Is Cannibalism a Myth?

> "Must have been someone he ate."
> —*Bob Hope, after hearing an*
> *African native burp loudly*
> *in* The Road to Zanzibar

Cultural anthropologists disagree over lots of things. Some still cling to the fading tradition of extreme cultural relativism, which forbids one to make value judgments about any culture. For example, it prevents an anthropologist from condemning Nazi anti-Semitism or American slavery because each was once embedded in a culture. It prevents condemnation of the widespread practice today, in Africa and elsewhere, of slicing off the clitoris of young women. More enlightened anthropologists find such relativism impossible to defend.

Some anthropologists, such as the late Margaret Mead, are strong believers in paranormal powers that are the stock in trade of parapsychologists. The vast majority of their colleagues consider this nonsense.

The latest, and perhaps the most bitter, of squabbles is over cannibalism. Most anthropologists remain convinced that cannibalism was widespread in the past and continues to flourish in obscure pockets of the world. This belief is defended in hundreds of papers and in such popular books as Eli Sagan's *Cannibalism* (1974) and Garry Hogg's *Cannibalism and Human Sacrifice* (1973).

A growing minority of anthropologists think otherwise. They are persuaded there is not now, nor has there ever been, a culture that routinely eats its dead, or that kills and devours its enemies. The debate simmered for decades until 1979, when it was blasted wide open by William Arens, an anthropologist at the State University of New York at Stony Brook. The blast was his explosive book *The Man-Eating Myth: Anthropology and Anthropophagy.* First published by Oxford University Press in 1979, it is currently available in softcover.

No one denies that during life-and-death emergencies, such as starving after a shipwreck or airplane crash, or during times of extreme famine, individuals may choose to eat human corpses rather than die.[1] No one denies there have been occasions among primitive peoples when, after a military victory, the body of a once-feared enemy leader was ritually devoured, either out of revenge or out of a belief that the enemy's powers would be acquired by the eaters. No one denies the existence of pathological serial killers, even in advanced cultures, who murder and feast on their victims.

The big question is this: Has cannibalism ever been a common custom? Arens's maverick opinion is that this kind of cannibalism is pure folklore, fabricated by the desire of one culture to feel superior to another. No trustworthy anthropologist, or any other reliable person, he contends, has ever directly observed a cannibal feast. The myth arose, he is convinced, from unsupported charges made by one culture against a hated neighboring culture, and was spread by gullible missionaries and naive anthropologists who swallowed every tall tale told them by friendly natives.

---

[1] "The Yarn of the *Nancy Bell,*" the best-known of W. S. Gilbert's comic ballads, is an account of cannibalism forced by a shipwreck. The yarn is told by the sole survivor, who ate nine of his shipmates.

A random check of current anthropology textbooks shows that most authors take for granted that cannibalism not only was widespread in the past, but also persists today among primitive tribes in Africa and South America and on islands in Oceania. Respected anthropologist Marvin Harris, in *Cannibals and Kings* (paperback, 1991), argues that cannibalism among the Mexican Aztecs was a way to obtain much-needed protein.

As for the public at large, who can doubt that there have been, and perhaps still are, fierce headhunters who eat their prey? Cannibalism continues to turn up in fiction and movies, in cartoons showing persons being boiled alive in enormous caldrons, and in crude jokes.

Have you heard about the cannibal whose cousin disagreed with him? Or the cannibal who admitted his wife made such a great pot roast that he would surely miss her? Or the Catholic cannibal of many years ago who on Friday ate only fishermen? Or the dentist who filled a cannibal's teeth while the cannibal was eating? Or the cannibal who told a psychiatrist he was fed up with people? Asked how he liked the psychiatrist he replied, "Delicious." A Charles Addams cartoon in *The New Yorker* in 1943 showed a cannibal mother taking her small son to a witch doctor and saying, "I'm worried about him, Doctor. He won't eat anybody." And we must not forget the cannibal who was walking through the jungle when he passed his mother-in-law.

Of several cannibal limericks I have come across, I like this anonymous one best:

> *A cannibal bold of Penzance*
> *Ate an uncle and two of his aunts,*
> *A cow and her calf,*
> *An ox and a half,*
> *And now he can't button his pants.*

The word *cannibal* derives from *Carib,* the name of an indigenous people of the West Indies and South America encountered by Columbus. In his journal Columbus described the Caribs as man-eaters. Why? Because he was told this by their neighbors, the Arawaks. How did Margaret Mead know that the Mundugumors of New Guinea were cannibals? The "gentle Arapesh" said so.

Arens documents hundreds of similar instances of one culture accusing another of cannibalism. The ancient Chinese thought Koreans were cannibals. Koreans thought the same about the Chinese. The Aztecs accused their Spanish conquerors of cannibalism. The conquistadors, who wrote all the books, are of course the main source for belief in Aztec cannibalism. When Arens was doing fieldwork in Tanzania, natives assured him that Europeans were cannibals. No culture in the world, he writes, ever claimed itself to be cannibalistic. Always, some other culture makes the accusation. As physicist Philip Morrison put it, praising Arens's book in *Scientific American* (September 1979), "[S]o it always seems: not us but them, or perhaps long ago and in another place."

Classic references to cannibalism go back to ancient Greek myths and the cyclops of Homer's *Odyssey.* Herodotus gave a secondhand account of cannibalism among the nomadic tribe called the Androphagi. In the Middle Ages, Catholics accused Jews of ritually devouring Christian babies. It was widely believed that witches also killed and gobbled babies. Incredibly, such a belief persists to this day among some ignorant fundamentalists obsessed by the notion that satanic rituals, including baby-eating, are taking place all over the United States!

Wild descriptions of cannibalism soared in the sixteenth century in hundreds of books by travelers and explorers. Hans Standen, a German sailor, wrote lurid accounts of savage headhunters among the Tupinamaba Indians of Brazil. His imaginary dialogues and appalling woodcuts were reprinted endlessly by other writers around the world.

Contributors to today's leading encyclopedias all defend ritual cannibalism: Ronald Berndt in the *Academic American Encyclopedia,* Paula Brown in Mircea Eliade's sixteen-volume *Encyclopedia of Religion,* and Leslie Spier in *Collier's Encyclopedia* and the *Encyclopedia Americana* (current editions). For horrendous earlier articles on cannibalism, see the eleventh edition of the *Encyclopaedia Britannica* (reprinted in the fourteenth). The current *Britannica* has no article on the topic. An even worse article (sixteen pages!) is by an erudite Anglican priest in the *Encyclopedia of Religion and Ethics.*

American Indians, especially the Iroquois tribes of New York, were branded cannibals by early white missionaries who saw them as cruel, mindless savages. The *Britannica*'s eleventh edition says that *Mohawk,* the name of an Iroquois tribe, probably means "man-eaters." There is not a shred of reliable evidence that the Iroquois, or any other native American culture, indulged in the ritual eating of human flesh.

Many anthropologists have written about cannibalism among Australian and New Zealand Aborigines. Michael Pickering, in *Cannibalism Among Aborigines?* (thesis completed in 1985 for the anthropology department of the Australian National University), concluded after an exhaustive study of the evidence that Aboriginal cannibalism never existed.

Descriptions of cannibalism are often mixed with unbelievable fantasy. Arens tells of an informant who, after disclosing to an anthropologist the human-flesh-eating customs of a neighboring people, added that the cannibals were all women who could, at will, turn themselves into birds!

The curious thing about the vast literature on cannibalism is the absence of firsthand accounts. Anthropologists never actually *see* a human-flesh-eating ritual. No photographs of the practice exist. "Cannibals are always with us," writes Arens, "but happily just beyond the possibility of direct observation."

The heated controversy between anthropologists who agree with Arens and those who dismiss his book as worthless is summarized as follows in Lawrence Osborne's excellent article "Does Man Eat Man?" in *Lingua Franca* (April/May 1997):

> The result of this uproar has been a crisis at the heart of the discipline, with different schools of anthropology—cultural, physical, and archaeological—turning in radically different verdicts on whether people are, or ever were, cannibals. It's enough to make some wonder if the always shaky alliances between anthropology's subfields are doomed to collapse altogether.

Mainline anthropologists have reacted to Arens's book with the same kind of fury they displayed toward Derek Freeman's *Margaret Mead and*

Samoa, a book exposing Mead's gullibility in taking at face value the myths told to her by Samoan pranksters. Anthropologists have yelled insults at Arens in meetings. They have pounded him relentlessly in their writings. Reviewers called his book "dangerous" and "malicious." The American Anthropological Association conducted a panel on cannibalism. Its attacks on Arens were published by the association in *The Ethnography of Cannibalism* (1983), edited by Paula Brown and Donald Tuzic.

In 1989, British anthropologist Sir Edmund Leach declared: "Montaigne writing about cannibalism in the sixteenth century is still far more convincing than Arens writing . . . in 1979." Anthropologist Vincent Crapanzano, reviewing Arens's book harshly in the *New York Times Book Review* (July 29, 1979), called it "poorly written, repetitive, snide."

Dr. Carlton Gajdusek, a physician who won a Nobel Prize for investigating a rare viral disease in New Guinea known as kuru, claimed the disease was transmitted by cannibals in the Fore tribe when they handled corpses. Believing the evidence to be so overwhelming for New Guinean cannibalism, he said, "It's beneath my dignity" to answer Arens. On the other hand, Lyle Steadman, of Arizona State University, after two years of work in New Guinea, failed to find any evidence of cannibalism.

In spite of his opponents, Arens's skepticism is slowly gaining respectability. Osborne reports in his *Lingua Franca* article that the 1996 *Encyclopedia of Anthropology,* in its entry on cannibalism, concludes: "Cannibals are largely creatures of our own surmise." Paul Behn, in the *Cambridge Encyclopedia of Human Evolution* (1992), writes: "Ritual or habitual cannibalism is rare or nonexistent: There are no reliable, firsthand witnesses of the practice, and almost all reports are based on hearsay." (See also Behn's article "Is Cannibalism Too Much to Swallow?" in *New Scientist,* April 27, 1991.) Noted anthropologist Ashley Montagu, long suspicious of cannibal accounts, has given Arens's book high praise.

In recent years paleontologists and archaeologists have claimed to have found evidence of cannibalism in human bone fragments among Neanderthal skeletons and in the burial sites of other ancient cultures. Archaeologist Timothy White, of the University of California at Berkeley, in his book *Prehistoric Cannibalism at Mancos* (1992), gives his reasons for believing that bone fragments in Pueblo Indian burial mounds in Colorado's Mancos Canyon prove that the Anasazi Indians had frequent cannibal feasts. He has likened Arens to a flat-earther who denies the world is round.

Not all experts buy White's conjectures. They find explanations other than man-eating for the conditions of his more than two thousand bone fragments. The fragments may indicate burial practices or eating by wild animals. Even if they are evidence of cannibalism, they could simply reflect isolated events brought on by famine. See archaeologist Behn's sharp criticism of White's book in *New Scientist* (April 11, 1992), and his contribution to the *Cambridge Encyclopedia of Human Evolution* previously mentioned.

Not being an anthropologist, I hesitate to take sides in this acrimonious controversy, though my sympathies at the moment are with Arens. Perhaps the truth is somewhere in between. Habitual cannibalism may be much rarer than believed by anthropologists who have a vested interest in rein-

One of the woodcuts in Hans Staden's *The True History and Description of a Country of Savages, a Naked and Terrible People, Eaters of Men's Flesh, Who Dwell in the New World Called America.* Staden is the fig-leafed man on the right, his hands clasped in prayer. (From Arens's book.)

forcing their earlier opinions. And cannibalism may not be as mythological as Arens supposes. Maybe time will settle the question.

Addendum

My column on cannibalism generated lots of mail.

Most of the letters, many too long to be printed in *Skeptical Inquirer,* strongly disagreed with Arens, and some expressed great disappointment with me for taking Arens seriously. Virologist D. Carleton Gajdusek's work in the 1960s with the Fore tribe of New Guinea was frequently cited as proof of persistent cannibalism. He argued that only cannibalism could explain the spread of a disease called kuru. No writer mentioned the contrary opinions of Lyle B. Steadman and Charles F. Merbe, of Arizona State University, in Tempe. In a paper in *American Anthropologist* they pointed out that Dr. Gajdusek and his associates never witnessed a cannibal feast firsthand, and that the spread of kuru could easily have been transmitted among the Fore by unsanitary mortuary practices.

Among recent books opposed to Arens's skepticism are Reay Tannahill's *Flesh and Blood: A History of the Cannibal Complex* (1975); Hans Askenasy's *Cannibalism: From Sacrifice to Survival* (1994); Frank Lestringant's *Cannibals* (1997); and Peggy Sunday's *Divine Hunger: Cannibalism as a Cultural System* (1986).

Arens replied to his critics in "Rethinking Anthropophagy," a lengthy paper in *Cannibalism and the Colonial World* (Cambridge University Press, 1998), edited by Francis Barker, Peter Hulme, and M. Iverson. There is special emphasis on the dubious claims of Dr. Gajdusek's study of the Fore tribe in New Guinea. Arens finds no reason to recant his original thesis, that although under certain conditions occasional cannibalism does exist, there is as yet no hard evidence that it is or has been a common practice in any culture.

Several readers sent me cannibal jokes. For example, a couple of cannibals are eating a stand-up comic. Says one of them, "Does this taste funny to you?"

An astronomer was in Africa to photograph a solar eclipse when he was captured by a tribe of cannibals. They intended to cook and eat him.

The astronomer recalled the episode in Mark Twain's *A Connecticut Yankee in King Arthur's Court*. He would tell the cannibal chief that powerful gods were protecting him. If they prepared to cook him, the gods would destroy the sun.

"Exactly when do you intend killing me?" the astronomer asked.

"At three this afternoon," the chief replied, "right after the total eclipse."

The limerick I quoted was rather tame. Julian Goldsmith, a University of Chicago geologist, sent me four improved third and fourth lines:

> *His mother-in-law*
> *(Though she stuck in his craw)*
>
> *A couple named Jones*
> *And her softer bones*
>
> *A circus fat lady*
> *And her sister Sadie*
>
> *Three visiting nuns*
> *and two of their sons*

CHAPTER 14

Alan Sokal's Hilarious Hoax

> It is simply a logical fallacy to go from the observation that
> science is a social process to the conclusion that the final
> product, our scientific theories, is what it is because of the social
> and historical forces acting in this process. A party of mountain
> climbers may argue over the best path to the peak, and these
> arguments may be conditioned by the history and social
> structure of the expedition, but in the end either they find a
> good path to the peak or they do not, and when they get there
> they know it. (No one would give a book about mountain
> climbing the title *Constructing Everest.*)
>
> —*Steven Weinberg,* Dreams of a Final Theory, *Chapter 7*

In a Spring/Summer 1996 issue devoted to what they called "Science Wars," the editors of *Social Text,* a leading journal of cultural studies, revealed themselves to be unbelievably foolish. They published an article titled "Transgressing the Boundaries: Toward a Transformative Hermeneutics of Quantum Gravity." It was written by Alan Sokal, a physicist at New York University. His paper included thirteen pages of impressive endnotes and nine pages of references.

Why were the editors foolish? Because Sokal's paper was a deliberate hoax, so obvious in its gibberish that any undergraduate in physics would have at once recognized it as a hilarious spoof. Did the editors bother to check with another physicist? They did not. To their everlasting embarrassment, at the same time they published the hoax, *Lingua Franca,* in its May/June 1996 issue, ran an article by Sokal in which he revealed the joke and explained why he had concocted it.

Sokal opened his parody with a strong attack on the belief that there is "an external world whose properties are independent of any human individual and indeed of humanity as a whole." Science, he continued, cannot establish genuine knowledge, even tentative knowledge, by using a "so-called" scientific method.

"Physical reality . . . is at bottom a social and linguistic construct," Sokal maintained in the next paragraph. In his *Lingua Franca* confessional he comments: "Not our *theories* of physical reality, mind you, but the reality itself. Fair enough: Anyone who believes the laws of physics are mere social conventions is invited to try transgressing those conventions from the windows of my apartment (I live on the twenty-first floor)."

Here are a few more absurdities defended in Sokal's magnificent spoof:

• Rupert Sheldrake's morphogenetic fields are at the "cutting edge" of quantum mechanics. (On Sheldrake's psychic fantasies see Chapter 15 of my *The New Age,* published by Prometheus Books in 1991.)

• Jacques Lacan's Freudian speculations have been confirmed by quantum theory.

• The axiom that two sets are identical if they have the same elements is a product of "nineteenth-century liberalism."

• The theory of quantum gravity has enormous political implications.

• Jacques Derrida's deconstructionist doctrines are supported by general relativity, Lacan's views are boosted by topology, and the opinions of Ms. Luce Irigaray, France's philosopher of feminism, are closely related to quantum gravity.

The funniest part of Sokal's paper is its conclusion that science must emancipate itself from classical mathematics before it can become a "concrete tool of progressive political praxis." Mathematical constants are mere social constructs. Even pi is not a fixed number but a culturally determined variable!

I hope no reader tries to defend this by pointing out that pi has different numerals when expressed in a different notation. To say that a notation alters pi is like saying 3 has a different value in France because it is called *trois*.

Pi is precisely defined within the formal system of Euclidian geometry, and has the same value inside the sun or on a planet in Andromeda. The fact that space-time is non-Euclidian has not the slightest effect on pi. African tribesmen may think pi equals 3, but that's a matter not of pure math but of applied math. This confusion of the certainty of mathematics within a formal system and the uncertainty of its applications to the world is a common mistake often made by ignorant sociologists.

The media had a field day with Sokal's hoax. Edward Rothstein's article in the *New York Times* (May 26, 1996) was titled "When Wry Hits Your Pi from a Real Sneaky Guy." Janny Scott's piece "Postmodern Gravity Deconstructed, Slyly" ran on the front page of the *New York Times* (May 18). Roger Kimball wrote in the *Wall Street Journal* on "A Painful Sting Within the Academic Hive." George Will, in his syndicated column, gloated over Sokal's flimflam. *Social Text*, he predicted, "will never again be called a 'learned journal.' "

The editors of *Social Text* were understandably furious. Stanley Aronowitz, cofounder of the journal, is a Marxist sociologist. He branded Sokal "ill-read and half-educated." Andrew Ross, another leftist and the editor responsible for putting together the special issue, said he and the other editors thought Sokal's piece "a little hokey" and "sophomoric." Why then did they publish it? Because they checked on Sokal and found he had good credentials as a scientist.

The strongest attack on the hoax came from Stanley Fish, an English professor at Duke University and executive director of the university's press, which publishes *Social Text*. Fish has long been under the spell of deconstructionism, an opaque and rapidly fading French movement that replaced existentialism as the latest French philosophical fad. In his "Professor Sokal's Bad Joke," on the *New York Times* Op-Ed page (May 21, 1996),

Fish vigorously denied that sociologists of science think there is no external world independent of observations. Only a fool would think that, he said. The sociologists contend nothing more than what observers *say* about the real world is "relative to their capacities, education, training, etc. It is not the world or its properties but the vocabulary in whose terms we know them that are [*sic*] socially constructed."

In plain language, Fish is telling us that of course there is a structured world "out there," with objective properties, but the way scientists *talk* about those properties is cultural. Could anything be more trivial? The way scientists talk obviously is part of culture. Everything humans do and say is part of culture.

Having admitted that a huge universe not made by us is out there, independent of our little minds, Fish then proceeds to blur the distinction between scientific truth and language by likening science to baseball! He grants that baseball involves objective facts, such as the distance from the pitcher's mound to home plate. Then he asks: "Are there balls and strikes in nature (if by nature you understand physical reality independent of human actors)?" Fish answers: No. Are balls and strikes social constructs? Yes.

Let's examine this more closely. The sense in which balls and strikes are defined by a culture is obvious. Chimpanzees and (most) Englishmen don't play baseball. Like the rules of chess and bridge, the rules of baseball are not part of nature. Who could disagree? Nor would Fish deny that pitched baseballs are "out there" as they travel objective paths to be declared balls and strikes by an umpire. Even the umpire is not needed. A camera hooked to a computer can do the job just as well or better. The basis for such decisions are of course cultural rules, but the ball's trajectory, and whether it goes over the plate within certain boundaries, is as much part of nature as the path of a comet that "strikes" Jupiter.

The deeper question that lies behind the above banalities is whether the rules of baseball are similar to or radically different from the rules of science. Clearly they are radically different. Like the rules of chess and bridge, the rules of baseball are made by humans. But rules of science are not. They are discovered by observation, reason, and experiment. Newton didn't invent his laws of gravity except in the obvious sense that he thought of them and wrote them down. Biologists didn't "construct" the DNA helix; they observed it. The orbit of Mars is not a social construction. Einstein did not make up $E=mc^2$ the way game rules are made up. To see rules of science as similar to baseball rules, traffic rules, or fashions in dress is to make a false analogy that leads nowhere.

It goes without saying that sociologists are not such idiots as to deny an outside world, just as it goes without saying that physicists are not so foolish as to deny that culture influences science. To cite a familiar example, culture can determine to a large extent what sort of research should be funded. And there are indeed fashions in science. The latest fashion in physics is the superstring theory of particles. It could be decades before experiments, not now possible, decide whether superstring theory is fruitful or a dead end. But that science moves inexorably closer to finding objective truth can only be denied by peculiar philosophers, naive literary critics, and misguided social scientists. The fantastic success of science in explaining and predicting, above all in making incredible advances in technology, is proof that scientists are steadily learning more and more about how the universe behaves.

The claims of science lie on a continuum between a probability of 1 (certainty) and a probability of 0 (certainly false), but thousands of its discoveries have been confirmed to a degree expressed by a decimal point followed by a string of nines. When theories become this strongly confirmed they turn into "facts," such as the fact that the earth is round and circles the sun, or that life evolved on a planet older than a million years.

The curious notion that "truth" does not mean "correspondence with reality," but nothing more than the successful passing of tests for truth, was dealt a death blow by Alfred Tarski's famous semantic definition of truth: "Snow is white" is true if and only if snow is white. The definition goes back to Aristotle. Most philosophers of the past, all scientists, and all ordinary people accept this definition of what they *mean* when they say something is true. It is denied only by a small minority of pragmatists who still buy John Dewey's obsolete epistemology.

Those who see science as mythology rather than an increasingly successful search for objective truth have been roughly grouped under the term "postmoderns." It includes the French deconstructionists, some old-fashioned Marxists, and a few angry feminists and Afrocentrists who think the history of science has been severely distorted by male and white chauvinism. Why did men study the dynamics of solids before they turned their attention to fluid dynamics? It is hard to believe, but one radical feminist claims it was because male sex organs become rigid, whereas fluids suggest menstrual blood and vaginal secretions!

A typical example of postmodern antirealism is Bruce Gregory's *Inventing Reality: Physics as Language*. The title tells it all. See my *Skeptical Inquirer* column "Relativism in Science" (Summer 1990), reprinted in *On the Wild Side* (Prometheus, 1992), for a review of this peculiar book. For a more resounding attack on such baloney, I highly recommend the recently published *Einstein, History, and Other Passions: The Rebellion Against Science at the End of the Twentieth Century* (Addison Wesley, 1996) by the distinguished Harvard physicist and science historian Gerald Holton.

The late Thomas Kuhn's famous book *The Structure of Scientific Revolutions* has been responsible for much postmodern mischief. Pragmatist Kuhn saw the history of science as a series of constantly shifting "paradigms." The final chapter of his book contains the following incredible statement: "We may, to be more precise, have to relinquish the notion, explicit or implicit, that changes of paradigm carry scientists and those who learn from them closer and closer to the truth." As if Copernicus did not get closer than Ptolemy, or Einstein closer than Newton, or quantum theory closer than earlier theories of matter! It takes only a glance at a working television set to see the absurdity of Kuhn's remark.

Fish and his friends are not that extreme in rejecting objective truth. Where they go wrong is in their overemphasis on how *heavily* culture influences science, and above all, in their obfuscatory style of writing. Examining interactions between cultures and the history of science is a worthwhile undertaking that may even come up someday with valuable new insights. So far it has had little to say that wasn't said earlier by Karl Mannheim and other sociologists of knowledge. Meanwhile, it would be good if postmoderns learned to speak clearly. Scientists and ordinary people talk in a language that takes for granted an external world with structures and laws not made by us. The language of science distinguishes sharply between language and science. The language of the sociologists of science blurs this commonsense distinction.

It is almost as if Fish were to astound everyone by declaring that fish are not part of nature but only cultural constructs. Pressed for clarification of such a bizarre view he would then clear the air by explaining that he wasn't referring to "real" fish out there in real water, but only to the *word* "fish." In a fundamental sense scientists and sociologists of science may not disagree. It's just that the sociologists and postmoderns talk funny. So funny that when Sokal talked even funnier in one of their journals they were unable to realize they had been had.

Lingua Franca, in its July/August 1996 issue, published an article by Bruce Robbins and Andrew Ross, coeditors of *Social Text,* in which they do their best to justify accepting Sokal's brilliant prank. Their reasons fail to mention the real one—their ignorance of physics.

In an amusing rejoinder, Sokal writes: "[M]y goal isn't to defend science from the barbarian hordes of lit crit (we'll survive just fine, thank you), but to defend the Left from a trendy segment of itself." His reply is followed by a raft of letters from scholars, some praising Sokal, some condemning him. They add little substance to the debate.

Addendum

Since Sokal's famous hoax appeared, so many books and papers have either favored or attacked postmodernism that it would take far too many pages merely to list significant references. Two very recent books are *A House Built on Sand: Exposing Postmodernist Myths About Science* (1998), edited by Noretta Koertge, and Ian Hacking's *The Social Construction of What?* (1999). Hacking, a Canadian philosopher of science, tries hard not to take sides.

Steven Weinberg's "Sokal's Hoax" in *The New York Review of Books* (August 8, 1996) is a brilliant demolition of postmodernism. *Free Inquiry,* in its Fall 1998 issue, ran an eighteen-page section on postmodernism featuring articles by sociobiologist E. O. Wilson and five other scholars, with wise final comments by philosopher John Searle. In the *New York Times* (November 11, 1998) Janny Scott reported on bitter ideological disputes over postmodernism at Duke University's English department that have caused the department to deconstruct. Stanley Fish was among those who departed. He is now dean of the College of Liberal Arts and Sciences at the University of Illinois in Chicago.

Alan Sokal's paper "What the *Social Text* Affair Does and Does Not Prove" appeared in *Critical Quarterly* (Vol. 40, Summer 1998, pp. 3–18). The previous year his book *Impostures intellectuelles,* written with Belgian physicist Jean Bricmont, was published in Paris, and simultaneously in England as *Intellectual Impostures.* A year later St. Martin's Press, under its Picador imprint, retitled its American edition *Fashionable Nonsense: Postmodern Intellectuals' Abuse of Science.* It omits a chapter on Henri Bergson's naive failure to understand relativity theory. The book swarms with juicy quotes of unbelievable absurdity and opaqueness. Here is a typical passage by Freudian ultrafeminist Luce Irigaray:

> Is $E=Mc^2$ a sexed equation? Perhaps it is. Let us make the hypothesis that it is insofar as it privileges the speed of light over other speeds that are vitally necessary to us. What seems to me to indicate the possibly sexed nature of the equation is not directly its uses by nuclear weapons, rather it is having privileged what goes the fastest . . .

Irigaray's incessant nonsense is topped by Jacques Lacan's crazy argument that the male sex organ is "equivalent to the square root of minus one."

Philosopher Thomas Nagel, reviewing the book for *The New Republic* (October 12, 1998), summed up France's cantankerous crew this way:

> The writers arraigned by Sokal and Bricmont use technical terms without knowing what they mean, refer to theories and formulas that they do not understand in the slightest, and invoke modern physics and mathematics in support of psychological, sociological, political, and

philosophical claims to which they have no relevance. It is not always easy to tell how much is due to invincible stupidity and how much to the desire to cow the audience with fraudulent displays of theoretical sophistication. Lacan and Baudrillard come across as complete charlatans, Irigaray as an idiot, Kristeva and Deleuze as a mixture of the two. But these are delicate judgments.

Of course, anyone can be guilty of this kind of thing, but there does seem to be something about the Parisian situation that is particularly hospitable to reckless verbosity.

The *Skeptical Inquirer* published in its March/April 1997 issue two angry letters from readers who accused me of unfair attacks on John Dewey and Thomas Kuhn. One of them, Daniel Nexon, was so furious that he canceled his subscription to the magazine as a protest against my "crossing over the line" from what he considered "acceptable discourse." I replied that of course my columns are polemical, and not intended to be polite technical arguments. I said that anyone interested in my views about truth and reality will find them carefully expressed in *The Whys of a Philosophical Scrivener* where I include a discussion of the famous conflict between Dewey and Bertrand Russell over how to define truth.

A few additional references:

The Flight from Science and Reason. Paul Gross, Norman Levitt, and Martin W. Lewis (eds). New York Academy of Sciences, 1996.

"When Science and Beliefs Collide." Janet Raloff, in *Science News,* Vol. 149, June 8, 1996, pp. 360–61.

"Postmodernism Exposed: Science Fiction." Peter Berkowitz, in *The New Republic,* July 1, 1996, pp. 15–16.

"Science as a Cultural Construct." Kurt Gottfried and Kenneth G. Wilson, in *Nature,* Vol. 10, 1997, pp. 545–47.

"The Sokal Hoax: At Whom Are We Laughing? Mara Beller, in *Physics Today,* September 1998, pp. 29–34.

CHAPTER 15

The Internet

A World Brain?

World Brain, one of H. G. Wells's many long-forgotten books, was published in 1938. Although written before the computer revolution, in many ways it anticipated the Internet and World Wide Web. But first, some background on Wells as a prophet.

In both science fiction and nonfiction, Wells's predictions were a fascinating mix of hits and misses. He took seriously the belief in canals on Mars, and intelligent Martians are featured in his novel *The War of the Worlds,* and in several short stories. In *Anticipations* (1901) he thought it unlikely that airplanes "will ever come into play as a serious modification of transport and communication." In the same book's chapter on modern warfare he wrote: "I must confess that my imagination, in spite even of spurring, refuses to see any sort of submarine doing anything but suffocate its crew and founder at sea." In *The Way the World Is Going* (1928) Wells anticipated the "complete disappearance of radio broadcasting."

In "The Queer Story of Brownlow's Newspaper," first published in *The Ladies' Home Journal,* April 1932, Wells described the contents of a 1972 newspaper. His hits include the use of color in newspapers, increasing space devoted to science news, a reduction in body clothing, and the collapse of Communism, although this did not happen until much later than Wells had expected.

The misses in Brownlow's newspaper far exceeded its hits. Wells thought that by 1972 nations would have been rendered obsolete by the rise of a world government. Heat from Earth's interior would replace fossil fuels. English spelling would be simplified, the stock market would vanish, the gorilla would be extinct, newspapers would be printed on paper made from aluminum, and a thirteen-month calendar would be adopted worldwide.

Wells failed to anticipate television even though Hugo Gernsback, who began reprinting Wells's science fiction in his *Amazing Stories,* was actually broadcasting television pictures in the twenties! One had to build a TV set to receive a postcard-size picture, but the technology was well underway.

Wells's last major effort to foresee the future was his imagined world history, *The Shape of Things to Come* (1932). Its misses were huge. Like Brownlow's newspaper, the book failed to foresee television, space flight, atomic energy, and computers.

Wells's wrong predictions were overshadowed by his single most astonishing hit. His novel *The World Set Free* (1914) opens with passages from the diary of a physicist who has split the atom and released atomic energy. The diary could have been written by Enrico Fermi. Wells's physicist agonizes over the horrendous results sure to follow from his achievement but he reasons that had he not made it, other scientists soon would have. The novel describes a world war started by Germany's invasion of France in the middle of the twentieth century. What Wells called "atomic bombs" are dropped from airplanes. The novel closes with visions of space explorations, beginning with trips to "that great silvery disk," the moon, "that must needs be man's first conquest of outer space."

World Brain, written while the clouds of World War II were gathering, consists of lectures and a few magazine articles. Wells saw knowledge increasing at an accelerating pace. At the same time, most people around the world remained incredibly ignorant. As Wells had earlier remarked, humanity is in a race between education and catastrophe. What could be done to raise the educational level of the world?

Humanity desperately needed, Wells was convinced, what he called a *Permanent World Encyclopaedia.* It would take the form of some forty enormous volumes that would be continually updated. At the time Wells wrote, specialized science journals were proliferating rapidly. Today there are some fifty thousand of them worldwide. What was needed so desperately, Wells maintained, was a central clearinghouse for this vast glut of information. The great encyclopedia would serve as a "world brain" by means of which information could be recorded and rapidly distributed around the world.

Wells likens the human race to intelligent persons with lesions in their brains—huge gaps between available information and public understanding. Horses have been replaced by cars, trains, and airplanes, Wells writes, but humanity is still in a horse-and-buggy stage. He emphasizes the rapid increase in travel and communication—what he calls the "abolition of distance." Yet in spite of such spectacular technical progress the world is like a ship in uncharted waters, sailing sluggishly toward a world community. For Wells, his mammoth encyclopedia would be a powerful force for unifying nations and speeding the coming of a war-free world.

Because English had become the world's most used language, Wells expected his encyclopedia to be in English. It would draw upon all the libraries of the world where information would be stored on microfilm:

It seems possible that in the near future we shall have microscopic libraries of record in which a photograph of every important book and document in the world will be stowed away and made available for the inspection of the student. . . . Cheap standardized projectors offer no difficulties. The bearing of this upon the material form of a World Encyclopaedia is obvious. . . . The time is close at hand when any student, in any part of

the world, will be able to sit with his projector in his own study at his or
her convenience to examine *any* book, *any* document, in an exact replica.

We can forgive Wells for not anticipating computers. Change "projector" to "computer" in the above passage and Wells's vision is surprisingly
accurate. Today a college student, even a high school student, is expected
to own a computer. At a few touches of the keys, pages from millions of
books and journals in libraries throughout the world can instantly flash on
a monitor. Copies of art in museums can be downloaded in full color. With
suitable equipment, music and voice recordings can be heard.

Research papers that once took years to come to the attention of scientists around the world are now starting to appear first as "e-prints" on the
Internet, that vast network of networks that allows computers around the
world to communicate with one another. Great efforts have been squandered in the past on research that duplicated work done years before but
which was unknown to the rediscoverers. Today, such wasteful duplications
are unlikely to occur because scientists can turn on search engines and
quickly track down previously published reports.

"This Encyclopaedic organization," Wells wrote, "need not be concen-
trated in one place; it might have the form of a network. It would central-
ize mentally but perhaps not physically. . . . It is its files and its conference
rooms which would be the core of its being, the essential Encyclopaedia.
It would constitute the material beginning of a real World Brain."

From this continually updated fount of information, Wells wrote,

would be drawn a series of textbooks and shorter reference encyclopae-
dias and encyclopaedic dictionaries for individual and casual use. That
crudely is the gist of what I am submitting to you. A double-faced organization, a perpetual digest and conference on the one hand and a system of publication and distribution on the other. It would be a clearing
house for universities and research institutions; it would play the role of
a cerebral cortex to these essential ganglia. On the one hand this organization should be in direct touch with all the original thought and research in the world; on the other it should extend its informing tentacles

to every intelligent individual in the community—the new world community.

Although Wells could not have known it at the time, he was writing about the Internet and the World Wide Web. How amazed and delighted he would have been by this revolution had he lived another half century!

Today the computer revolution is changing the world in ways as hard to predict now as it was hard to predict the consequences of the industrial revolution. Almost every conceivable topic now has one or more websites. Nearly anyone, of any age, race, sex, or intelligence, can start a World Wide Web page. The 1998 edition of Luckman's massive *World Wide Web Yellow Pages* lists ten thousand websites, a mere fraction of the total number that grows larger every hour. Are you interested in learning more about any famous person? Chances are high that you will find Web pages devoted to that person. Popular science magazines, weekly news magazines, leading newspapers, and news television shows such as CNN and C-Span now have their websites. A dozen sites are devoted to cigars alone! Every major religion and bizarre little cult is online. Thousands of literary classics can be downloaded. Entire encyclopedias are available as well as entire runs of certain magazines such as *National Geographic.* More and more libraries, including the Library of Congress, are putting their card catalogs online.

Consider my own hobby, conjuring. At last count there were almost two hundred websites devoted to such topics as magic history, card tricks, magic organizations, and magic periodicals. Almost fifty are sponsored by magic equipment dealers. More than a hundred are run by individual magicians!

Stores of all sorts are going online to sell goods, especially new and rare books, that you can purchase without leaving your home. You can buy groceries, a car, an airplane ticket, even browse through flea markets on the Web. Businesses are starting to decentralize into "virtual offices" where employees can work at home. Fiber optics will soon be replacing copper wires, allowing thousands of messages to be transmitted simultaneously over one fiber line. Cyberspace is only in its infancy!

Were Wells alive today I think he would be writing about the Net's good and bad features. He would, of course, be enthusiastic about how the Net is speeding communication among scientists and scholars. He would welcome the way it is bringing diverse cultures closer together into a kind of global village of residents less inclined to slaughter one another.

On the other hand, I fancy that Wells would also deplore the darker corners of cyberspace. The incredible amount of valid information available to anyone online is not easily distinguished from the equally incredible smog of junk science, sleaze, and nonsense. At the moment, the Net is in a state of anarchy with almost no government controls. Do we need such controls? If so, how far should they go? At the moment, anyone can say anything on the Web. Cyberspace is infested with idiots, con artists, and purveyors of cybersmut. Advertising, especially the annoying ads that pop up in corners of the monitor, increasingly contaminates computer screens. As someone has said, it's like opening your mailbox and finding one letter, two bills, and sixty thousand pieces of junk mail. :-(

Stand-up comic Jackie Mason in a recent one-man Broadway show had these comments about netheads who boast of being able to converse with strangers:

> You want to talk to people all over the world? People don't talk to the guy next door. . . . A guy calls you up and he's got the wrong number. Do you start a conversation? . . . Last week a guy tells me: "I spoke to a chap from Siberia, a mountain climber from Siberia." . . . If a mountain climber from Siberia came over to your house and said, "Hello, I'm a mountain climber!" would you say, "Come in, I'm dying to talk to you! All my life I wanted to talk to a mountain climber from Siberia!" :-)

Anyone online can turn on a search engine to contact a thousand websites devoted to pseudoscience, the paranormal, and the occult. Yahoo! lists multiple sites on biorhythm, alchemy, ghosts, astral projection, crop circles, dowsing, spontaneous human combustion, the hollow earth, witchcraft, voodoo, palmistry, sea monsters, and hundreds of other outrageous topics. Countless sites are devoted to UFOs and parapsychology as well. Astrology alone has some 150 sites!

Hundreds of sites deal with every variety of alternative medicine: iridology (diagnosing ills from spots on the iris), crystal healing, therapeutic touch, naturopathy, aromatherapy, urine therapy, homeopathy, orgone therapy, and on and on. Scientists may have learned how to use the Web wisely to keep up with global research, but the average browser is hard put to filter out the trash, not to mention the hardcore pornography that the government doesn't yet know how to curb.

Lost among the thousands of sites devoted to bogus science and the occult are a minuscule number of sites run by skeptics. Here are a few main ones:

CSICOP/*Skeptical Inquirer:*
www.csicop.org

Australian Skeptics:
www.skeptics.com.au

James Randi Foundation:
www.randi.org

National Center for Science Education on evolution versus creationism:
www.natcenscied.org

Skeptic magazine:
www.skeptic.com

Quackwatch, Stephen Barren's pages on health frauds:
www.quackwatch.com

Robert Carroll's Skeptic's Dictionary:
www.skepdic.com

News of interest to skeptics:
www.skepticnews.com

Prometheus Books:
www.prometheusbooks.com

Also on the Net's downside is the increasing number of computer own-
ers suffering from a new addiction comparable to alcoholism and com-
pulsive gambling. Millions of netomaniacs spend all their spare time surfing
the Net, participating in chat rooms (there now are some twenty thousand
of them!), checking news groups, playing computer games, and corresponding with persons whose character and even age they may not know.
Elderly men and women find themselves in erotic dialogue with teenagers.
Youngsters may be induced to meet child molesters. Unsuspecting browsers
may not be aware when they are victims of hoaxes and practical jokes.

What Wells liked to call humanity's current "age of confusion" is certainly reflected in the vast confusion of the Web. It will surely be many
decades before the Internet settles down, if it ever does, into a healthy, ad-
mirable world brain, a force for good that Wells hoped would hasten the
coming of a saner world.

Addendum

Wells's *World Brain* was reprinted in paperback in England in 1994, with
the new title *H. G. Wells On the Future of World Education*. Peter Lonsdale,
treasurer of England's H. G. Wells Society, sent me a copy of his review of
this book as it appeared in the society's *H. G. Wells Newsletter*.

Illustrator/author Ron Miller wrote to tell me about "A Logic Named
Joe," a short story by Murray Leinster (pen name of science fiction author
William Fitzgerald Jenkins) that appeared in *Astounding Science Fiction*
(March 1946). The story contains an amazingly accurate forecast of the In-
ternet, a worldwide network linking together what Leinster calls "logics,"
but today would be called desktop computers.

Every hour the World Wide Web grows larger and more chaotic. At the
time I write (fall 1999) there are some 800 million pages on the Web, far
too many to be listed with current search engines. A recent study shows

that the engines are capable of indexing only about 16 percent of the Web. If a new website appears, it takes the most popular search engines an average of six months to list it. Because websites reporting science information constitute only about 6 percent of the Web, it is conceivable that a search engine devoted exclusively to science could keep track of this information, but for now there is no such engine.

In 1999 Dover reprinted Wells's *Anticipations* in a paperback edition. In my introduction I discuss Wells's many hits and many misses.

CHAPTER 16

Carlos Castaneda and
New Age Anthropology

Among American anthropologists a raucous minority believe firmly in the reality of ESP, PK (psychokinesis), precognition, and other psychic wonders, especially in the paranormal powers of the shamans or sorcerers of primitive cultures. In March 1999 a section of the American Anthropological Association, calling itself the Society for the Anthropology of Consciousness, sponsored a five-day conference at the University of California at Berkeley. It was its nineteenth spring conference.

I did not attend, nor have I sent $140 for tapes of the fifty lectures by counterculture speakers, but I do have a copy of the meeting's twenty-seven-page program sent to me by my friend Jim Breese. "A few issues back," he said in a letter, "you deplored some current agenda at Temple

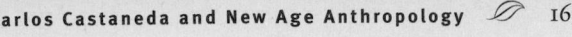

University. Well, if you think that's bad, have a look at the enclosed program."

I was indeed appalled by what I saw. What startled me most was the number of speakers who enthusiastically praised the late bogus anthropologist Carlos Castaneda.

Not much is accurately known about Carlos because he was known to fabricate information about his life. He so successfully evaded efforts to photograph him that the only photo published in the United States that shows his face clearly was taken at a college graduation in 1959. It is now known that he was born in 1925 as Carlos Cesar Arana Castaneda, in Cajamarca, Peru (not in Brazil as he often claimed). A few years after finishing high school in Lima, he married, but deserted his wife and child to go to California in 1951. For a few years he lived in San Francisco, then moved to Los Angeles, where he took courses in journalism and creative writing at Los Angeles City College. In 1960 he married Margaret Runyon. They lived together for only six months, but were not divorced until thirteen years later.

A photograph of Margaret accompanies her article "My Husband Carlos Castaneda," in *Fate* (February 1975). She describes him as five feet five with dark eyes, curly black hair, and thighs and legs "disproportionately short." His face was round and cherubic, his skin nut-brown.

Carlos had a habit of suddenly vanishing, to return unexpectedly without telling Margaret where he had been. Phone calls were always made from public booths. There were periods, she writes, when he drove a taxi, kept accounts for a ladies' apparel shop, and clerked in a liquor store, from which he would bring her "wonderful wines." These jobs may or may not have been real because it was never possible to know when Carlos was telling the truth. Letters to his wife were unsigned.

In 1968, Carlos was an anthropology student at the University of California at Los Angeles (UCLA) when the university's press published his first book, *The Teachings of Don Juan: A Yaqui Way of Knowledge*. The book hit the New Age market like a bombshell. Sales went through the roof and Castaneda became instantly rich and famous.

The book tells how Castaneda, on a field trip to Mexico in 1960, met Don Juan at a dreary bus depot in Arizona. He turned out to be an elderly

Yaqui sorcerer with vast magical powers and a habit of giggling even more often than the Maharishi of Transcendental Meditation. Castaneda became Don Juan's apprentice. His progress involved the frequent taking of such drugs as peyote cactus, Jimson weed, and various hallucinogenic mushrooms. The book's main theme is that beyond our ordinary world is an extraordinary realm in which one can talk to animals, even become animals, and experience all sorts of wonderful miracles. This other world, so familiar to Yaqui shamans, is just as real as this one.

Castaneda followed his first book with nine others, all best-sellers, which were translated into some twenty languages. In order of publication they are *A Separate Reality: Further Conversations with Don Juan; Journey to Ixtlan: The Lessons of Don Juan; Tales of Power; The Second Ring of Power; The Eagle's Gift; The Fire from Within; The Power of Silence; Further Lessons of Don Juan;* and *The Art of Dreaming.*

In 1972, UCLA, in a fit of self-deception, gave Carlos a doctorate in anthropology. His thesis was based on *Journey to Ixtlan.* Although the public, smitten by New Age fantasies and the pleasures of mind-altering drugs, gobbled up Castaneda's books, mainstream anthropologists were outraged. Careful investigations found his books riddled with contradictions, outright errors, and rafts of material pilfered from other authors. Don Juan existed only in Carlos's imagination. As sociologist Marcello Truzzi was the first to say, Castaneda's books were the greatest science hoax since the Piltdown Man. Irate anthropologists demanded that the university withdraw Castaneda's Ph.D. It refused. Professor Walter Goldschmidt, then chairman of the university's anthropology department—who wrote a fulsome forward to Castaneda's first book—said in 1978: "We possess no information that would support the charges. . . . I am not going to say *mea culpa.*"

In 1999 the University of California Press, motivated by shameless greed, issued the thirtieth anniversary edition of Carlos's first book. They promoted it not as a work of pseudo-anthropology but as a classic still relevant to readers longing to escape from the dull world of ordinary reality to a magical land of Oz.

I don't have the space to relate the thousands of paranormal events described in Castaneda's ten fantasies. I will mention only the funniest. In his first book Carlos tells how Don Juan, high on drugs, transformed him into a crow. Mind you, this was no hallucination. Carlos became a real, live crow. When Don Juan tossed him into the air, he flew away. The sorcerer told Carlos that three crows would signal his death and that after he died he would be reincarnated as a crow.

It is hard to believe that some anthropologists still consider Castaneda to have been a serious, competent researcher into the stupendous wonders of Yaqui Indians. Here are glimpses at some of the abstracts of papers delivered at the March meeting of the counterculture anthropologists.

Amy Smith, of Salt Lake City, Utah, spoke on "A Castaneda Way of Knowledge: Implications of an Anthropological Legacy." She claims that Castaneda's efforts in "personally exploring nonordinary states of consciousness, addressing the existence of multiple realities and other unusual phenomena, using both emic and etic interpretations, and documenting and reporting these experiences through a narrative ethnography were ground breaking achievements that remain essential to the field."

Edith Turner, of the University of Virginia, in her paper on "The Teachings of Castaneda," sees his research as a great liberation. "He has taken us—like Dante—through a dark passage out the other side into a state of enlightenment." He has freed us from "capitalism, communism, consumerism, Church rationality and exclusivism; from reductionism, fundamentalist Marxist scientism and its type of elitism, which condemns the · folk; and lastly from nihilism, the eternal war of revenge upon the sins of many societies which can never be forgiven." Not only that, but "we are allowed to talk with empathy, for instance, about the Nigerian sacred Bori personage, about the Hindu guru, the Brazilian *mae de santo,* Rumi, the Tibetan oracle, Jesus, the Bal Shem Tov, the Dalai Lama, Black Elk, and at last, Don Juan." Whew!

Roy Wagner, also at the University of Virginia, calls his paper "Consciousness Is that Part of Consciousness of Which We Are Conscious: How the Ancient Seers and Shamans of Mexico Short-Circuited the Energy-Body." I fail to grasp what he means when he adds: "For having devastated the 'inner' (subjective, imaginal, hence 'heuristic') support for our ability to conceptualize things in the world, Don Juan has turned the whole 'subjectivity' fantasy inside-out: we do not 'think' his lessons except insofar as they think us."

Michael J. Winkelman, of Arizona State University, spoke on "Epistemological Perspectives on Castaneda." His abstract reads in full:

Castaneda's concepts such as "Separate Reality," "stopping the world," and "tonal and nagual" are analyzed from an epistemological perspective. His approaches to the sorcerer's practice were explicitly epistemological, centrally concerned with processes through which human conceptual processes are structured and contribute to the reality known. Central aspects of Castaneda's training involved learning how to suspend the ordinary epistemological constructs and enter into a "natural epistemological mode." These innate constructs of nature and the human brain/mind are revealed by comparing them to similar epistemological systems and practices found in other contemplative disciplines. The nature of the sorcerer's epistemological development is assessed from the perspectives of genetic epistemology.

Other abstracts of papers given at the conference are written in the same murky, mind-numbing jargon. Here are a few samples.

"Insights from anthropological linguistics include indigenous knowledge of telepathy in natural communication, and discovering human languages of wholeness. . . ." (Dan Moonhawk Alford, California State University.)

"In children with MPD [multiple personality disorder], the Original Personality goes for safekeeping from Physicalspace into Thoughtspace, where it is cared for by wise and loving collections of intelligent energy." (Ralph Allison, of Los Osos, California.)

"Cultural manifestations of human consciousness are the result of both immanent and transcendent aspects of a principle of action stemming from a philosophical tension inherent in the nature of awareness." (Richard L. Amoroso, of the Noetic Institute.)

"The culturally diverse delineation of these experiences will be considered while discussing the posited referencing of the extrasensory experiences in question within the context of an energetic-wave world view." (Renate Dohman, of Goldsmith College, London.)

The conference's keynote address, on "The Varieties of Dissociative Experience," was given by Stanley Krippner, a well-known parapsychologist. Here is how the meeting's program described his speech:

> Stanley Krippner will describe a cross-cultural, postmodern, transpersonal model of dissociation based on the work of Ruth-Inge Heinze and Rhea White. Dissociation is contrasted with flow, identification with the ego-self is contrasted with identification with the All-Self, control is contrasted with loss of control. Each of these examples of dissociation and integration is evaluated on the basis of whether it is socially constructive or destructive. Examples will include (among others) dissociative identity disorder, "channeling," shamanic journeying, and acting. First person accounts will be cited; for example, Robert Louis Stevenson, Edgar Bergen, Shirley MacLaine, Chris Sizemore ("Three Faces of Eve"), Maria Sabina, and J. Z. Knight (who "channels" Ramtha).

Unbounded praise of Castaneda by New Age anthropologists can be found in many books. The wildest reference is *Extrasensory Ecology: Parapsychology and Anthropology* (1977), a collection of papers edited by Joseph K. Long. (See Richard de Mille's biting review in *Skeptical Inquirer*, Spring/Summer 1978, pp. 108–12.) The book's first paper, by Long, is about Castaneda. He calls Castaneda's first book "one of the most important books in the field of anthropology."

The book's second paper is a vigorous attack on Castaneda by Agehananda Bharati, but most contributors to this bizarre anthropology are great admirers of Castaneda. Margaret Mead, for instance, writes: "Carlos Castaneda has developed a method which makes the American Indian religious experience available to non-anthropologists who would never be able to get the same experience from rereading interlinear texts."

William S. Lyons's paper defends the ability of psychics to "see" things in nonordinary reality, such as seeing human auras. "If, for example, the director of the National Science Foundation could 'see' what Don Juan claims to 'see,' psychical research would be more than likely to be as readily funded as body language research." Long, commenting on this paper, suggests that the pineal gland, or "third eye," is the "point of focus" in the kind of "seeing" described by Castaneda. In a paper on the evolution of psi, Long defends Castaneda's accounts of the psychic abilities of certain animals and birds. On page 261 Long calls Shirley MacLaine an "erudite amateur anthropologist."

Long's ludicrous book must be read to be believed. Jule Eisenbud, who for a decade tried to persuade the world that Ted Serios could project his thoughts onto Polaroid film, writes on "Perspectives in Anthropology and Parapsychology." Long thinks Eisenbud's book on Serios is "one of the best documented cases ever presented for a regular PK effect." Physicist Evan Harris Walker, writing on "The Compleat Quantum Mechanical Anthropologist," assures us that quantum effects explain Uri Geller's ability to bend spoons.

Two leading proponents of psychic archaeology, J. N. Emerson and Jeffrey Goodman, each contribute to Long's book. Goodman tells how a psychic in Oregon went into trances and remote-viewed a spot in Flagstaff, Arizona. The accuracy of his visions enabled Goodman to discover deeply buried artifacts.

Goodman believes that the Garden of Eden was not in Africa but in California. He denies that American Indians came here across the Bering land bridge from Asia. They went the other way, California to Asia! From California they migrated to Africa and other parts of the world. And where did these California Indians come from? In Goodman's first book, *Psychic Archaeology: Time Machine to the Past* (1977), a map shows how they came, from the lost continents of Atlantis and Lemuria!

Goodman's second book, *The Genesis Mystery: A Startling New Theory of Outside Intervention in the Development of Man* (1983), rehashes his earlier book but adds surprising new material. He claims that Darwin stole all his ideas from Alfred Wallace. Although our bodies evolved from lower animals, a sharp transition took place when God planted human souls in bestial bodies. Goodman seems unaware that he is defending current Roman Catholic opinion.[1]

Carlos Castaneda died in Westwood, California, in 1998. "His only real sorcery," writes Kathryn Lindskoog in her entertaining book *Fakes, Frauds, and Other Malarky* (1993), "was turning the University of California into an ass." The next time you come close to a crow, try calling out "Hello Carlos!" If you are high enough on peyote, you might hear the bird answer.

Time, March 5, 1973 (Time/Life Syndication)

[1]See Kenneth Feder's article "American Disingenuous: Goodman's 'American Genesis'— a new chapter in 'Cult' Archaeology," *Skeptical Inquirer*, Summer 1983, pp. 36–48; J. R. Cole's review of Goodman's *Psychic Archaeology* in *Skeptical Inquirer*, Spring/Summer 1978, pp. 105–8; and Cole's review of *The Genesis Mystery* in *American Antiquity*, Vol. 50, 1985, pp. 692–93.

References

"Don Juan and the Sorcerer's Apprentice." Unsigned cover story in *Time*, March 5, 1973, pp. 36–45.

"On Don Juan's Last Laugh." A review of Castaneda's *Tales of Power*, by Joyce Carol Oates, in *Psychology Today*, September 1974, pp. 10, 12, 130.

Castaneda's Journey: The Power and the Allegory. Richard de Mille. Capra Press, 1976.

Seeing Castaneda: Reactions to the "Don Juan" Writings of Carlos Castaneda. Daniel Noel (ed.). Putnam's, 1976.

"Does Don Juan Live on Campus?" Paul Preus, in *Human Behavior*, November 1978, pp. 53–57.

The Don Juan Papers: Further Castaneda Controversies. Richard de Mille (ed.). Ross-Erickson, 1980.

"The Sorcerer's Birthday: The Fiction of Carlos Castaneda." Gregory McNamee, in *The Bloomsbury Review*, September/October 1988, p. 31.

Carlos Castaneda, *Academic Opportunism, and the Psychedelic Sixties.* Jay Courtney Fikes. Millennia Press, 1993.

"What Hath Carlos Wrought?" Robert McGrath, in *The Skeptic*, March/April 1993, pp. 11–12.

"Portrait of a Sorcerer." Keith Thompson, in *New Age Journal*, March/April 1994, pp. 152–53.

"Carlos Castaneda and Don Juan." Jay C. Fikes, in *Encyclopedia of the Paranormal*, Gordon Stein (ed.). Prometheus, 1996.

"Shaman or Sham?" Hal Cohen, in *Lingua Franca*, September 1998, pp. 22–24.

Addendum

Since this column was published I obtained a subscription to *Anthropology of Consciousness,* a journal published three times a year by SAC (Society for the Anthropology of Consciousness), a section of the American Anthropological Association. According to this journal's inside front cover, the society's primary areas of interest are altered states of consciousness, religion, possesion, trance, dissociative states, shamans, mediums, mystics, psychic phenomena, and psychic archaeology.

The first issue I received (Vol. 10, June/September 1999) is devoted mainly to traditional shamans and what are called neo-shamans. Throughout, the authors distinguish between what they call experimental, experiential, and theoretic approaches to consciousness phenomena. Their papers include rafts of footnotes and long bibliographies, and all are written in a peculiar kind of jargon that I find hilarious. For example, here is how Ina Rösing, of Germany's University of Ulm, ends her paper on "Lies and Amnesia in Anthropological Research," a paper based on her field research of Andean medicine men and East Tibetan shamans:

Regarding the multiple problems of any transcultural hermeneutics, the present discussion does perhaps imply only a small twist of the viewing angle. It turns the view from the apparently grand to the modest attitude of the lowered gaze, directed to the "refuse" and the "side products" of research, to "lies," imagination, and the seepage waste of amnesia.

Part VII

UFOs

CHAPTER 17

Claiborne Pell,
Senator from Outer Space

pell inhis heaventh glike noughty times ∞,
—*James Joyce*, Finnegans Wake

Claiborne De Borda Pell, senior United States senator from Rhode Island and former chairman of the Senate's powerful Foreign Relations Committee, announced in 1985 that he will not run for reelection in 1996. For thirty-five years "Wellborn Pell," as colleagues sometimes call him because of his wealthy father, has been the most strident member of Congress in trying to persuade government agencies to increase funding for psychic research.

Interest in psi phenomena and other New Age folderol has always been part of the circus inside the Beltway. The Pentagon, the Central Intelligence Agency, and the Federal Bureau of Investigation all have people strongly supportive of psi funding. Both the Army and Navy have sponsored such research, costing taxpayers millions of dollars. Usually the work has been top secret, listed under code names that conceal the nature of the investigations. The secrecy is due in part to fear of ridicule by skeptics, and especially by Christian fundamentalists who suspect the agencies are in league with Satan.

In 1984 the Army Research Institute, fearful that the Soviets were decades ahead of the United States in paranormal research, funded an investigation by the National Academy of Sciences. Psychologist Ray Hyman was placed in charge of a subcommittee to report on the status of parapsychology. The study concluded there is no good evidence that psi phenomena exist. Some of the psi research conducted by CIA officials is hard to believe, the academy found. The CIA had tried training psychics to look at photos of Soviet cars and tell what was going on inside them. The officials considered seriously the technique of puncturing tires by sticking pins into photographs!

A full report was published in 1987 by the National Academy Press under the title *Enhancing Human Performance.* The report was of course roundly blasted by parapsychologists and by Pell as a misuse of government funds.

U.S. News & World Report (December 5, 1988), in an article titled "The Twilight Zone in Washington," estimated that "one-fourth of the members of Congress are actively interested in psi, be that healing, prophecy, remote viewing, or physical manifestations of psychic powers." Texas Democrat Jim Wright, former House speaker, said he believes he has strong psi abilities to see future events. We all remember how former White House residents Ronald and Nancy Reagan were devout believers in astrology. Dates of the president's important meetings were scheduled by Joan Quigley, their San Francisco astrologer. In my opinion, however, no one in Washington has rivaled Senator Pell in combining ignorance of science with extreme gullibility toward the performances of psychics.

Born in New York City in 1918, and a graduate of Princeton University, Pell has a record of being one of the Senate's most liberal Democrats. Although an Episcopalian, he is strongly pro-choice on abortion rights, a brave stance considering Rhode Island's large Roman Catholic constituency. His pro-labor record is consistent. He has been awarded almost forty honorary degrees. Other honors include Italy's Grand Cross of the Order of Merit and France's Legion of Honor. He was one of the founders of the National Endowment for the Arts.

Pell's efforts to combat environmental pollution led him in 1988 to introduce a bill for government funding of a New Age organization named the National Committee on Human Resources. Senators Albert Gore and Nancy Kassebaum were cosponsors. The committee was to have included two members "with training and experience in extraordinary performance results," a euphemistic way to describe parapsychologists. Ridiculed by other senators as a "spoon-bending bill," it quickly died. As one congressman put it, "The giggle factor is off the meter on this one."

Pell buys almost everything on the paranormal scene. His office shelves are jammed with books on the paranormal, including the many autobiographies of Shirley MacLaine.

Pell is on the advisory board of the International Association of Near-Death Studies—studies purporting to establish that persons close to death often get peeks into the hereafter. He is also on the board of the Institute of Noetic Sciences, an organization devoted to psi research.

In 1987 Pell invited Uri Geller, the self-proclaimed psychic, to Capitol Hill to demonstrate his alleged powers in an electronically bugproof room. Hanging on his office walls Pell has a spoon bent by Geller, a framed photo of Geller, and drawings of a "smiley face" made by Pell alongside a duplicate made by Geller, supposedly using ESP.

Pell admits that on occasion Geller may use magic. "Geller was a magician when he was younger," Pell told a reporter. "Maybe when his intuitive processes fail, he can back them up with sleight of hand." This is a commonplace remark made by psi researchers whenever a medium or psychic is caught cheating.

In the late 1980s the magician James Randi was in Washington to receive an award for excellence in public speaking. The award was presented by fellow conjuror Harry Blackstone, Jr. Sitting in the audience, lean and frail, was Pell. When he watched Randi bend a spoon until it broke, Pell became visibly agitated. One of Pell's associates took the two spoon pieces to Pell, who carefully wrapped them in a handkerchief. After the ceremony

Pell visited Randi backstage. Pell was angry because Randi had presented the spoon-bending as a magic trick. Pell had seen Geller perform this feat and I believe Pell was absolutely persuaded it could only be done by psychic means.

Pell challenged Randi to duplicate a drawing the way Geller had done. Pell produced a pad and a pen. While drawing a figure on the pad, Pell explained that he knew all about pencil reading, and would hold the pad in such a way that Randi could not see the pen wiggling. Pell also said he was aware that an impression of the drawing could be made on the pad's second sheet. He ripped off the top sheet, folded it twice, and put it in his pocket without handing Randi the pad.

Randi found a piece of paper and drew on it. He folded the sheet and placed it under Pell's foot. "If I duplicated what you drew," said Randi, "will you admit I have done it by trickery?"

Pell bent over to pick up and open Randi's paper. He turned pale and visibly trembled when he saw that Randi had exactly duplicated his drawing of an equilateral triangle.

Randi has given me permission to explain how he did this. In the act of tearing the sheet from the pad, Pell had allowed Randi an upside-down glimpse of what he had drawn! Blackstone had also seen the triangle, and was doing his best to keep from laughing. So much for Pell's ability to test a psychic! Was he convinced that Randi had performed a magic trick? Not on your life. His unflappable comment later was "I think maybe Randi is a psychic and doesn't realize it." Several naive parapsychologists have come to similar conclusions about Randi, evidently believing themselves too smart to be fooled.

For seven years Pell had on his staff, at a reported $50,000 a year, one of the nation's top promoters of the paranormal, Cecil B. Scott Jones. A handsome white-haired man, Jones was for forty years a Navy pilot and intelligence officer. He still retains top secret security clearance. For several years he taught political science at two Wyoming institutions: Casper College and the University of Wyoming at Laramie.

While a Navy attaché in India, Jones said, he had a paranormal experience so shattering that he told a reporter he could not describe it for fear of embarrassing the government. He added that this event "enabled me to do my intelligence assignment with much greater speed than one ordinarily expected." (See C. Eugene Emery, Jr.'s, article "Fear of Ridicule the Main Roadblock: Pell Aide Likens Government to Ostrich When It Comes to Psychic Phenomena," in the *Providence Sunday Journal,* July 17, 1988.)

After Jones left the Navy and obtained a doctorate in international studies at American University in 1975, his apparent obsession with the paranormal steadily grew. He tried unsuccessfully to persuade some companies to market a technique for telepathic communication. As he told Emery, "The venture failed because the companies and potential customers in the government were afraid of ridicule."

Jones has been a believer in UFOs ever since he saw a silvery disk in the skies when he was a Korean War fighter pilot. His one book, *Phoenix in the Labyrinth,* was published by his Human Potential Foundation in 1995. It consists of six speeches he gave at UFO conferences from 1988 through 1994.

Jones said he is certain that for decades Earth has been visited by extraterrestrials. He dislikes calling them aliens, preferring the term Visiting Others. He has no idea who they are or whether they come from distant regions of our universe, from parallel worlds in higher dimensions, or from the future. He said he is persuaded that the executive branch of our government, as well as top Russian officials, has information about the Visiting Others, which it is keeping from Congress and the public. "If I knew what was going on," he added, "I'm not sure I would tell you." For years he has urged the executive branch to stop its campaign of secrecy, disinformation, and cover-up. In one speech Jones said that if the president ever reveals what he (the president) knows, Congress would call for his impeachment.

"The government . . . may have painted itself into a corner," he said in 1988, "and after some forty-odd years the paint is still wet. . . . It is the nation's and the global best interest to assume that if direct ET encounters have not yet taken place, it is not too soon to anticipate such encounters and to make sensitive preparations for them."

Jones's craziest lecture, given in Denver in 1992 at a symposium on UFO research, concerned the great explosion that occurred on June 30, 1908, in Tunguska, Siberia. Astronomers agree that Earth was struck by a comet or huge meteorite. Not Jones. He said he believes it was a crashed UFO, cylindrical in shape.

Many pages in his book are devoted to the efforts of five psychics employed by Psi Tech, a commercial firm that claims to have perfected sophisticated techniques for remote viewing, not only of distant places, but also of past events! Jones published these psychics' Tunguska research results, complete with their crude sketches of what the Tunguska crashed object looked like. The pictures are in wild disagreement. Some of the psychics saw the object as unmanned, and called it "self-conscious." Others described it as manned by humanoids. All five agreed, however, that it came from a very distant world of sentient beings, perhaps from another dimension, entering our space-time through a wormhole. One psychic, who saw the object as egg-shaped, said it came from the future.

In 1984, Jones invited Pell to a seminar organized by parapsychologists. A self-proclaimed believer in psi since his college days, Pell was convinced by the seminar, he said, that it was essential for the Senate to have someone in a position to persuade the government to take psi phenomena more seriously. In 1985, Pell hired Jones as his aide.

On one occasion Jones and a psychic visited an aquarium in Texas and tried to communicate by telepathy with a dolphin. The results were inconclusive. Jones has since suggested that dolphins could be used to locate the remains of flying saucers that have crashed into the sea. On another occasion Jones sponsored an effort by psychic mediums to contact dead Soviet bigwigs and urge them to beam thoughts of peace to living Soviet leaders. In 1986 Jones invited Pentagon officials to his home to hear taped spirit voices, one supposedly from William Randolph Hearst.

Jones's tireless efforts to persuade the government to fund paranormal research hit its most preposterous level in October 1990. Jones wrote to then secretary of defense Richard Cheney to say that a group of parapsychologists had made a truly amazing discovery. In listening to speeches related to the Persian Gulf War made by President George Bush, Secretary of State James Baker III, and Cheney himself, a mysterious word had turned up when the tapes were played backward. The word was "simone."

"I mention this," Jones wrote, "in case it is a code word that would not be in the national interest to be known." The speculation was that the sub-conscious minds of the speakers were inadvertently introducing the secret code word into their reversed speech! Jones's boss, Pell, admitted that "while it sounds wacky, there might be some merit" to it because he respected "Scott's [Jones's] responsible role in life." (See John Diamond's Associated Press account of October 20, 1990; C. Eugene Emery, Jr.'s, article "Pell Aide Hears Code in Backwards Speeches," in *Skeptical Inquirer,* Summer 1991; and a report in *Harper's Magazine,* January 1991, p. 25.)

It turned out that the source of Jones's warning came from David Oates, an electronics enthusiast from Australia, then living in Dallas. Oates had coauthored a 1987 book in Australia on reverse speech therapy, whatever that is. Oates told Emery that when he played backward the Persian Gulf War speeches of the three political leaders he heard "simone" five times. Jones thought the repetition of "simone" was significant enough to send a warning to Cheney about the possibility of unintentionally disclosing a top secret code word.

Emery wrote:

When I asked Oates for an example of a "simone" message, he cited an August 8, 1990, news conference by Bush, where the backwards phrase that caught his interest was "Simone in the sands." He played the tape for me. Like most backwards messages, the phrase isn't very clear, but you can hear it if you're told what to listen for.

What were listeners actually hearing when Bush said "Simone in the sands"? Oates said the president's words were "Iraq has massed an enor-mous war machine." "Simone," he said, came from the sounds in the word *enormous.*

Columnist Philip Terzian said the affair reminded him of the Beatles' white album, in which listeners heard the backwards message "I bury Paul." "I've been reading [Jones's] letter to Richard Cheney backwards," Terzian wrote, "and [I] am certain I can hear a voice say: 'I bury Pell.' "

Oates told Emery he had been researching backward speech for seven years and considered his discovery that subconscious thoughts emerge in reverse speech to be one of "tremendous value." The backward phrases are of course highly symbolic and have to be carefully interpreted. For example, the phrase "I am Sir Lancelot" turned up on one of Oates's backward tapes. It signified, he said, that the speaker subconsciously thought of himself as a knight or savior. Oates said he moved to the United States in the hope of getting academic recognition for his great work.

When he first heard "simone" in Bush's speech, Oates said, he thought it might refer to a friend or relative, but after hearing it again in Gulf crisis speeches by Baker and Cheney, he began to suspect it was a secret code word. The phrase that emerged in Baker's backward speech was "simone won't shine."

At the time of this letter from Jones to Cheney, Pell was running for reelection against Congresswoman Claudine Schneider. She was unable to capitalize on the "simone" flap, Emery wrote, because she too is a believer in the paranormal. Pell easily won reelection. Uri Geller took partial credit for this victory. "I will beam my energy to him to win," he had told a reporter. (See the *Denver Post,* February 2, 1990, and "The Flip Side of Simone Is Enormous," by C. Eugene Emery, Jr., in the *Providence Journal-Bulletin,* February 10, 1990.)

Jones left his job with Pell in 1991 to devote himself full-time as president of his Human Potential Foundation. Its offices are in Falls Church, Virginia, where Jones now lives. The foundation was originally financed by Laurance Spelman Rockefeller. Pell serves on its board. From 1985 until recently, Jones had been a trustee of the American Society for Psychical Research, serving as its president from 1989 until 1992.

It is not known what Pell plans to do after retiring. In reporting his decision not to run again, *Time* (September 18, 1995) headed its article "Senator Oddball." For decades, said *Time,* Pell has been "Capitol Hill's most eccentric denizen," an inhabitant of the "Pell zone." Among many Pellisms is his way of saying "formal greetings" when he meets someone, and "too peachy" to describe a flowery speech. His clothes are baggy tweeds, he seldom shaves, and he tends to mumble when in doubt. As a tribute to his wealthy father, he wears his father's belt, which is so long that he has to wrap it twice around his waist. He even wears it jogging. Among his nicknames, said *Time,* are Stillborn Pell, Wellborn Pell, Senator Magoo, and the Senator from Outer Space.

When he planned one of his annual parties for staffers, Pell tried to borrow two camels from tobacco heiress Doris Duke, who also lived in Newport. He wanted the camels to graze on his lawn as a bizarre special attraction. Doris dissuaded him on the grounds that camels like to spit at people they don't know.

Another member of Congress who is tireless in promoting psi is Rep. Charles Grandison Rose III from North Carolina. He has been in Congress since 1973 and has served since 1977 on the House Select Committee on Intelligence. Like Pell, Rose said he firmly believes the military should spend much more money developing weapons that use psi powers—weapons that could make the old explosives obsolete.

Rose founded the Congressional Clearinghouse on the Future to give psychics a chance to address political leaders in Washington. He has advocated a government-funded "psychic Manhattan Project" to develop clairvoyant and psychokinetic techniques for foiling enemies. But Charlie Rose's career is another story.

Addendum

In late November 1995, the Defense Intelligence Agency disclosed the existence of its top secret program code-named Stargate, which was declassified and suspended in spring 1995. Over twenty years, $20 million was spent on the program that included studying six psychics who claimed to have powers of clairvoyance, called "remote viewing," that was supposedly used for spying. The CIA, which monitored Stargate, decided on the basis of reports by psychologist Ray Hyman and others that remote viewing was useless for intelligence work and no more public funds should be wasted on such research.

From 1985, Stargate was directed by Edwin May. His star performer was former Army intelligence officer Joe McMoneagle, who now runs a company, Intuitive Intelligence Applications, with his astrologer wife. They charge clients $1,500 a day.

On November 28, 1995, Ted Koppel's *Nightline* program on ABC-TV interviewed May, former CIA director Robert Gates, statistician Jessica Utts, a psi researcher introduced only as Norm, and Hyman. Gates said the CIA monitored Stargate only because the Russians were doing similar research, and because of pressure from a few unnamed congressmen. The results of Stargate's research were of no value, he said, and no CIA decisions were based on them.

May, Utts, and Norm all stoutly defended Stargate as validating remote viewing. Hyman, the token skeptic, was allowed only a few seconds to say that in his opinion remote viewing remains unconfirmed.

Senator Claiborne Pell (United States Senate Historical Office)

CHAPTER 18

Courtney Brown's
Preposterous Farsight

Cosmic Voyage: A Scientific Discovery of Extraterrestrials Visiting Earth. By Courtney Brown. Dutton, 1996. 275 pp. $23.95.

My first reaction to this preposterous book was "It's a hoax." But no, Courtney Brown actually exists. An associate professor of political science at Emory University, in Atlanta, Georgia, he has written several books about social science and one on mathematics titled *Chaos and Catastrophe Theory.* A "Ph.D." follows his name on the front of the jacket.

On the back cover are two laudatory blurbs. One is by Whitley Strieber, author of many books about UFO abductions, including his own abduction. The other is by Harvard psychiatrist John Mack. Mack firmly believes that aliens from higher space-time dimensions are visiting Earth and taking humans aboard their spacecraft to perform on them unspeakable operations. His 1994 book *Abduction,* published by Scribner's, was a best-seller. Harvard can't get rid of him because he has tenure.

Brown says that after TM (Transcendental Meditation) training—his wife is a TM instructor—he mastered the advanced TM-Sidhi Program, which teaches "Yogic flying" and other sidhis such as making oneself invisible and walking through walls. He highly recommends two books by the Maharishi. Brown next attended the Monroe Institute, in Faber, Virginia, where he learned telepathy and attained a high level of consciousness. He urges readers to obtain three books by Robert Monroe.

Brown's final level of training was by an expert in SRV (scientific remote viewing), a recent term for clairvoyance. He claims to have acquired not only the ability to remote-view distant spots on Earth, but also the abilities to see remote parts of the universe and witness historical events both in the distant past and in the far future. Readers who want to develop similar powers are urged to contact Brown at his Farsight Institute, Box 49243, in Atlanta. For $3,000 you can take a one-week course in remote viewing, with follow-up courses for additional fees.

Cosmic Voyage is a record of staggering "facts" Brown uncovered in more than thirty SRV sessions under the supervision of a man he calls his "monitor" or "instructor." Brown never names him, but he is known to be Ed Dames, a retired Army major, now president of a psychic research organization called Psi Tech, based in Beverly Hills, California.

Before Brown became Dames's pupil, Dames was on record as an ardent believer in a Galactic Federation of aliens who roam the galaxy in advanced spacecraft. He believes that Martians are living in New Mexico, that there are alien sites on the moon, and that the stone "face" on Mars was carved by a Martian civilization. For a while he was associated with the ill-fated ten-year research project Stargate, sponsored by the CIA to investigate possible military uses of remote viewing. After spending some $20 million, the CIA closed down the project as a waste of taxpayer money.

Brown's remote viewing sessions, supervised by Dames, enlarged greatly on Dames's beliefs. Critics of remote viewing point out that the easiest way results can be contaminated is by what is called "front loading." This means that the monitor, or the subject, or sometimes both know in advance what the "target" is. In Brown's sessions, Dames always knew the target. The possibility of conscious or unconscious influence over the subject, by comments and leading questions, is obvious. Because Brown's sessions were frontloaded, even parapsychologists who believe in remote viewing consider his sessions worthless.

According to Brown, millions of years ago a race of hairless humanoids lived on Mars. They had big eyes, light skin, and telepathic abilities. A wandering comet or maybe an asteroid grazed the planet, severely damaging its atmosphere and rendering the red planet unfit for habitation. The Martians were forced to move into underground caverns, where most of them still are today.

A Galactic Federation of superbeings dispatched to Mars a rescue team of altruistic humanoids called the Greys (because of their color). The Greys had earlier been forced to abandon their planet after carelessly allowing its environment to degenerate, much as we are now allowing ours. Their culture's collapse was facilitated by a mysterious evil dictator who suffered from "low self-esteem" and may have been none other than the Biblical Lucifer.

Before the Greys lost their homeland, they lived mainly on fish. Their sex drive was then much stronger than ours, although their "genitals were quite small by human standards." The Greys are short, with heads shaped like a praying mantis's and huge black eyes. They communicate by telepathy and have a lifespan of two hundred years.

These benevolent superbeings have spacecraft capable of going galactic distances at speeds faster than light and of moving back and forth in time. They can alter matter, permitting their ships to fly right through mountains, just the way they did in Spielberg's movie *Close Encounters*. The Greys are slowly evolving into still higher beings and eventually will "merge with God."

In recent decades several hundred Martians have been transported by the Greys to a spot north of Santa Fe, New Mexico, where they live in caverns beneath the mountain called Santa Fe Baldy. Their technology is 150 years ahead of ours. They are no longer bald. Thanks to their genetic engineering they are beginning to look like us. They have their own spaceships, which soon will be helping the Greys to bring more Martians to

Earth. Indeed, the base at Santa Fe Baldy will eventually become an immigration processing center for Martian refugees.

Another group of Martians has been taken by the Greys to an unidentified village in South America, where they live disguised as native Indians. The Greys were able to alter Martian genes so they can withstand Earth's stronger gravity and breathe our atmosphere. Both the Martians and the Greys are much concerned over how rapidly we are destroying our environment. They are eager to help us mend our ways but are unwilling to contact us until we make the first move of letting them know our desire to contact them.

Brown somehow discovered *The Urantia Book,* a mammoth tome purporting to be written by celestials under the supervision and editing of Dr. William Sadler, a well-known Chicago psychiatrist who died in 1969. Brown buys *The Urantia Book*'s claim that living among us, doing their best to aid us, are invisible beings called Midwayers (because they are on a plane midway between us and angels). "The discovery that the Midwayers actually exist," Brown writes, "was a shock that reverberated through the consciousness of the military's SRV team for quite a few years."

Brown believes that *The Urantia Book* is mostly accurate in its cosmology and its elaborate hierarchy of billions of higher entities and gods, including Jesus. His main objection to *The Urantia Book* is its failure to recognize reincarnation. Brown considers us composite beings. Our physical body grows old and dies, but a spiritual entity or "soul"—Brown persists in calling it our "subspace"—is immortal. After the human body dies, the soul travels upward forever, as *The Urantia Book* teaches, to inhabit more advanced bodies on other worlds. Thus Brown and *The Urantia Book* agree on endless reincarnation after death but disagree on our having had incarnations before birth. (For more than you may care to know about the Urantia movement, see my *Urantia: The Great Cult Mystery,* published by Prometheus Books in 1995.) At the moment, Urantians are divided over whether Brown's book is good publicity for their bible, or bad publicity because of Brown's other wild beliefs.

In remote-viewing sessions with Dames, Brown visits both Martians and Greys many times, entering their minds to learn their deepest secrets and motives. He interviews a translucent Jesus whose hair seems made of light. Jesus is friendly and has a great sense of humor. Brown also visits the Buddha and Guru Dev, who was the Maharishi's mentor. He remote-views President Clinton in the Oval Office. Dames tells Brown, "I could have had you go into his head, but that would have been an invasion of privacy." I found this the funniest line in Brown's book.

Brown moves back millions of years to view the desolation of Mars after it was damaged by the comet or asteroid. He goes forward three hundred years to witness the tragic plight of humans on Earth after our environment has hopelessly deteriorated. He contacts members of the Galactic Federation. He visits a world in the Pleiades star cluster, where he sees bewildered Americans who have been taken there by the Greys to preserve their genetic stock. The planet has two suns, one large and yellow, the other a smaller white dwarf. Brown believes some UFOs may be piloted by humans from our future. "It sounds weird," he told a reporter, "but I suppose you could be watching a ship fly by and you, as a future human, could be in it."

Brown contacts Adam and Eve. He finds that the history of this pair, as told in *The Urantia Book,* is essentially correct. They were were not the first humans on earth, but genetic engineers sent to earth by supermortals to supervise a "genetic-uplift" of our race. The uplift was the seeding of the planet with a new species that would evolve into humans. Gene manipulations of Earth life by superior beings have been going on for millions of years, Brown reveals. This explains what biologists call "punctuated equilibrium." Brown follows *The Urantia Book* in taking this to mean that new species appear "suddenly" in one generation, rather than evolving slowly by a series of mutations.

Brown remote-views the Civil War's Battle of Gettysburg. "It would be well worth the efforts," he advises, "for historians to revisit the battle using SRV." Imagine how all our history books will be rewritten after historians master the art of remote-viewing the past! Brown's clairvoyant visions are punctuated by his constant exclamations of "Gosh!" and "Wow!" as they continue to astonish him, especially after Dames reveals the nature of a target.

Our hopes of surviving the dark days ahead depend on how soon we decide to meet and cooperate with the Martians and Greys. Brown has not the slightest doubt that our government knows all about these aliens and their frequent abductions of humans so they can experiment on us to find out how to improve our genes. He is convinced, so help me, that the Greys have been invading the sleeping minds of writers who produce scripts for *Star Trek: The Next Generation.* These writers are totally unaware that their clever ideas are designed by the Greys to accustom Earthlings to the reality of extraterrestrials eager to transform our culture.

The time is growing short. Brown urges our government to abandon its stupid policy of secrecy and openly seek contacts with the Greys and Martians.

The only earlier book about UFOs I can think of that is nuttier than this one is George Adamski's *Inside the Space Ships* (1955). As an abductee aboard a UFO, Adamski saw bustling cities on the far side of the moon. Adamski, of course, was a charlatan, whereas Brown actually believes what he writes. Both books read as if they were efforts at science fiction written by a ten-year-old.

Emory University must be enormously embarrassed by having Brown on its faculty. As in Harvard's problem with Dr. Mack, they can't fire Brown because he has tenure, and Emory's president believes in academic freedom for its teachers.

In a recent interview in the *Kansas City Star,* Brown said that if NASA's planned Martian probes fail to show evidence of a Martian civilization, his career in higher education will be kaput. The 1994 probe that vanished a few days before it was to orbit the red planet to take pictures, Brown is convinced, was shot down by the Martians, who don't want to be observed!

"All the prestige that I've got is resting on whether there is anything in this," Brown told the reporter. "I'd be crazy if I went public with something like this without being certain about what's going on. . . . I'd be dead as an academic. I couldn't even get a letter published in Dear Abby."

It is a sad story of an intelligent, sincere man who has turned himself into a gullible dunce. It will be amusing to see how he reacts a few years from now, after the Martian probes find no traces of an advanced Martian civilization.

Addendum

Scott Lilienfeld's article "The Courtney Brown Affair and Academic Free-dom" appeared in the May/June 1997 issue of *Skeptical Inquirer.* Lilienfeld, a psychologist at Emory, tells how he challenged Brown to a simple test of his ability to remote-view objects in an adjacent room. Brown indignantly refused. "The tests you talk about are very old-hat," he said in his e-mail response, adding that the current status of remote viewing "goes light-years beyond that which your letter suggests."

Accompanying Lilienfeld's article is a statement by William Chace, Emory's president, stating that although he does not agree with Brown's "non-Emory activities," he defends Brown's right to pursue them. Of course Brown has such a right, but the damage to Emory's reputation has been immense. An editor of *Emory Wheel,* the university's student newspaper, urged action against Brown for "sullying the university's good name."

In November 1996, Brown played a major role in what can now be called the great Hale-Bopp flap. It all began in 1995 when Alan Hale and Thomas Bopp sighted a comet heading our way. It became known as the Hale-Bopp comet. Art Bell, the late-night radio host whose shtick is interviewing bogus scientists, reported that the comet was on collision course with earth. When this proved to be false, Bell came up with an even wilder possibil-ity. He interviewed an amateur astronomer, Chuck Shramek, who had posted on the Internet a photograph of Hale-Bopp that revealed a bright "Saturn-like object," four times the size of Earth, trailing the comet. Could it be a giant spaceship?

Brown, a friend of Bell's, got into the act by appearing twice on Bell's show. He said he had asked three of his Farsight Institute's crack remote viewers to take a look at the mysterious object. As Robert Sheaffer reported (*Skeptical Inquirer,* March/April 1997), one of the psychics saw the object as "a large, dense, magnetic object, powerful, ominous, and centrifugal." A second viewer found the thing to be a climate-controlled spaceship. This was confirmed by the third expert, who described the ship as "hard, smooth, and rounded."

Alan Hale examined Shramek's photograph. The fancied spaceship turned out to be the star SAO 141894, distorted by refraction. (See Hale's article "Hale-Bopp Comet Madness," in *Skeptical Inquirer,* March/April 1997.) Although the ship did not exist, it played a crucial role in the horrendous mass suicides of the Heaven's Gate cult, as will be told in the next chapter.

On April 11, 1997, Brown posted on his website a paranoid message to the U.S. government. It had come to his attention, he wrote, that a group inside the government was planning a terrorist attack on his Farsight Institute. "We have been in direct contact with an outside group that has offered us both guidance and protection. (From herein, we will refer to this group as our 'Friends.') We believe that you already know who this group is. What you do not know is that this group may not be able to protect us further should the government of the United States make an overt decision at the highest level to abandon us." Who are these Friends? They are none other than the alien "unseen friends" of the Urantia cult!

Brown goes on to say that on April 4, 1997,

we activated our new technology transfer protocols. We remote viewed an extraterrestrial artifact that is in the possession of the United States government. The fact that this object was in your possession was told to us by our Friends. They informed us only that it was rectangular in shape, its approximate size, that it was ET technology, and that the government did not know the purpose or operation of this object. Using our new technology transfer protocols (which, we admit, were partially given to us by our Friends), we targeted this object and discerned its purpose, how it works, etc. We are not reporting all of our findings here because it has come to our attention that such devices are common to all ships operated by our Friends, and that other governments also have such devices, or know about them. We want to limit our initial dissemination of our findings to the U.S. government for now. This will require a visit to our facilities in Atlanta by an official governmental representative to review our data.

We expect and hope that these findings will be useful to you. We are capable of offering you greater detail should you want or require it. Also, our technology transfer protocols, while already quite extensive, will be expanded dramatically shortly to enable us to tackle a wide range of very difficult problems. Our Friends want you to know that they are willing to work with you using the means of consciousness. They want to help you, but they want you to meet them halfway, by allowing us to develop the technologies of consciousness that allow for an intimate interaction between themselves and humans. We have not been informed of any limit to the help which they are willing to offer as long as we receive it using means with which they are comfortable.

Not only is a group in the U.S. government planning to demolish the Farsight Institute, but Brown says he has information from the Friends about a planned terrorist attack on New York City using a tactical nuclear weapon stolen from the former Soviet Union. This attack is imminent, and the government must act quickly to prevent it. Brown again begs the government to visit his facilities in Atlanta, and "rather than trying to destroy us, consider protecting us."

I am often bashed for ridiculing extreme forms of pseudoscience rather than treating it as if it were serious speculative research. Physicist Jeremy Bernstein, writing about this in his book *Science Observed*, defends the ridicule technique:

Quantum mechanics is *not* Zen Buddhism. Photons do *not* display manifestations of consciousness. Relativity theory has nothing to do with ethical relativism. Creationism is not a rival scientific theory of the origin of species. Evolution is *not* a speculation, and so on. If people read popular science with misguided expectations, in the long run this will manifest itself in a loss of popular support for, and interest in, real scientific research.

For this reason, among others, I believe that a scientist such as myself who writes for the general public has both an opportunity and a responsibility to call attention to nonsense when he comes across it. On the other hand, to write about nonsense and save the writing from degenerating into tedious polemic requires thought. Here is where the humor comes in. It is, I think, a great deal more effective and certainly a lot more fun to try humorously to make pseudoscientific notions appear as ludicrous as, in fact, they are. In doing so, one runs the risk of being taken lightly even when one's intention is not necessarily to be funny. That is a risk I am willing to assume.

Courtney Brown continues to be an enormous embarrassment to Emory University. In 1999 Dutton published his *Cosmic Explorers,* a sequel to *Cosmic Voyage.*

Brown's ego and self-deception are monumental. He is totally unaware that once his early remote viewing established a wild science fiction scenario about ETs, his later remote viewing would confirm the earlier results. *Cosmic Explorers* opens with detailed instructions on how to go about practicing SRV (Scientific Remote Viewing), a psi ability formerly called clairvoyance. It is an ability Brown believes anyone can acquire if he or she tries hard enough. Learning how to meditate before a session is a big help.

The rest of Brown's book is an account of his own SRV efforts to find out what is going on in our galaxy. The Martians, who look much like us, are continuing to escape from their underground caverns on Mars to be transported to Earth by the Greys. They now live inside Santa Fe Baldy, a mountain in New Mexico. How come you can't go to Mount Baldy and discover the entrances? Because there *are* no entrances! The advanced technology of the Greys allows them to make their huge spaceships invisible, then go right through the mountain's side! They solidify on the other side in a large hangar where the Martians, aided by some humans, are building new spaceships.

The Greys are incredibly intelligent, highly telepathic, generous, and kind. They want to cooperate with our government to save Earth from environmental disasters. They love us, but don't want to force us to do anything. Our government has to make the first move.

Outside Earth, in galactic subspace ("subspace" is Brown's term for a space in another dimension), a tremendous battle is now underway between the beautiful, spiritual Greys and a fierce tribe of ugly extraterrestrials.

Brown calls them Reptilians because they have an orange-green reptilelike skin. They are fighting the Greys for control of Earth. They are our enemies. They intend to breed with humans to create a new race of hybrids. Already there are Reptilians on Earth, where they live inside at least one mountain that Brown leaves unspecified. They also have a command center in space, between Earth and the moon. We can't see it because it is invisible. Brown isn't sure whether the Reptilians are here from a distant past or from a distant future. They are the ETs who abduct Earthlings in their spaceships and perform upon them unspeakable acts.

In one of his remote-viewing sessions Brown actually viewed God himself. He discovered that in the distant past God self-destructed into fragments and that he is now in the process of reuniting the fragments and restoring himself. We and all other forms of life are those fragments. Our bodies are perishable, but each of us has an immortal soul. Brown argues convincingly that if we don't have souls independent of our body, how could we travel to distant realms to remote-view them, or travel to scenes back in time or far into the future?

There is in outer space, or rather in subspace, a vast Galactic Federation which does its best to oversee the galaxy in a manner similar to how our United Nations tries to stop conflicts on Earth. In *Cosmic Voyage,* Brown tells of remote-viewing the Buddha. In this book he tells of once again viewing Buddha. Buddha is now a leader in the Galactic Federation. He is greatly upset over the war between the Greys and the Reptilians, and is trying hard to help resolve the conflict.

Major Ed Dames, who runs Psi Tech, was Brown's unnamed mentor in *Cosmic Voyage.* He is nowhere mentioned in *Cosmic Explorers.* It may be that he and Brown, now heading rival remote-viewing institutes, have fallen out. Dames, like Brown, is a frequent guest on Art Bell's radio talk show. *Time,* in an article on Bell (August 9, 1999), reported that in December 1998 Dames predicted that terrorists would unleash biological weapons in July, in Shea or Yankee Stadium, and that he was "locked in a psychic battle with Satan."

I closed *Cosmic Explorers,* which reads like primitive science fiction, with a feeling of sadness for the president of Emory, who is unable to think of a way to fire Brown, and even greater sadness for Brown's wife and son. What do *they* think of his fantasies, I wonder?

Brown's Farsight Institute, where he trains persons in remote viewing, may be reached at P.O. Box 49243, Atlanta, GA 30359. You can keep abreast of the institute's research by calling up its website, www.farsight.org. For information about Brown, contact his personal website, www.court neybrown.com.

CHAPTER 19

Heaven's Gate

The UFO Cult of Bo and Peep

For there are some eunuchs, which are so born from their
mother's womb; and there are some eunuchs, which were made
eunuchs of men; and there be eunuchs which have made
themselves eunuchs for the kingdom of heaven's sake. He that is
able to receive it, let him receive it.

—*Jesus, Matthew 19:12*

The nation's shocked reaction to the suicide in March 1997 of thirty-eight happy brainwashed innocents and their demented leader, at Rancho Santa Fe, California, has been twofold. The event has reawakened public awareness of the enormous power of charismatic gurus over the minds of cult followers, and it has focused attention on the extent to which the myth of alien spacecraft has become the dominant delusion of our times. A recent *Newsweek* poll revealed that almost half of Americans believe UFOs are for real and that our government knows it. As if a secret this monumental could be kept by our political leaders for more than a few hours!

Rumors about space aliens and their snatching of humans show no signs of abating. Harvard psychiatrist John Mack has published a book about the abductions of his patients. The rumors are magnified mightily by endless other books, lurid movies, and shameless radio and television shows. Ed Dames, who runs Psi Tech, a psychic research center in Beverly Hills, California, was the first to proclaim that his "remote viewers" had spotted a massive spacecraft trailing Comet Hale-Bopp.

Dames's claim was "confirmed" by three psychics at the Farsight Institute in Atlanta, headed by Courtney Brown, Dames's former pupil. Brown teaches political science at Emory University. He is as embarrassing to Emory as Mack is to Harvard. The previous chapter reviews Brown's *Cosmic Voyage,* a crazy book telling how aliens are being shuttled to Earth to live under a mountain near Santa Fe, New Mexico.

Who was most responsible for the Rancho Santa Fe horrors? They were two neurotic, self-deluded occultists: Marshall Herff Applewhite and his platonic companion Bonnie Lu Trousdale Nettles. Their story reads like bad science fiction.

Born in Spur, Texas, in 1931, the son of a Presbyterian minister, Applewhite graduated as a philosophy major at Austin (Texas) College in 1952. He had brief stays in seminary school and in the Army Signal Corps. But he was gifted with good looks and a fine baritone voice, and his chosen career was in singing and music. He received a master's degree in music from the University of Colorado, starring in numerous operas produced in Houston and in Boulder while pursuing his degree. Throughout his musical career, he held a number of teaching positions and conducted numerous church choirs.

For several years in the sixties, Applewhite taught music at St. Thomas University, a small Catholic school in Houston. The university fired him in 1970 over an affair with a male student. Struggling to control his homosexual impulses, depressed, and hearing voices, he checked into a psychiatric hospital in 1971. He told his sister he had suffered a heart attack and had had a near-death experience.

It was in this hospital that Applewhite's life took its fateful turn. His registered nurse, Bonnie Nettles[1] (she was forty-four, he forty), was a former Baptist, then deep into occultism, theosophy, astrology, and reincarnation. Somehow she managed to convince Applewhite that they were aliens from a higher level of reality who had known each other in previous earthly incarnations. In the coming months and years, they would develop their bizarre religion, believing they had been sent to Earth to warn humanity that our civilization was about to collapse as foretold in Revelation, to be replaced by a new one after the battle of Armageddon and the destruction of Lucifer. They believed that Lucifer (a cut or two below Satan), aided by his "Luciferians," had long controlled our planet. It is, in fact, Lucifer's demons who have been piloting those spaceships that are abducting humans.

How can one escape the coming holocaust? Not by being "raptured," as Protestant fundamentalists teach, but by being beamed up to spacecraft operated by benign superbeings and taken to the gates of heaven. (Judging by the recent tragedy at Rancho Santa Fe, if you are male, the best way to make this journey is to cut off your testicles, then kill yourself!)

Shortly after their meeting, and fired with the divine mandate to rescue as many humans as possible from the destruction of our world as we know it, Applewhite and Nettles embarked on their mission, Nettles abandoning a husband and four children in the process. (Applewhite, a father of two, was already divorced.) The pair quickly became inseparable in a strange bond that psychiatrists call the "insanity of two." It develops when two neurotic persons live together and reinforce each other's delusions.

Indeed, Applewhite and Nettles began calling themselves The Two. They came to believe they were the "two witnesses" described in Chapter 11 of Revelation. Verse 7 predicts that when the two witnesses "finish their testimony" they will be murdered. After three and a half days, God will res-

---

1. Some sources say that Applewhite met Nettles when he was visiting a friend in the hospital. Nettles's daughter says they met at a drama school. (See *New York Times,* April 28, 1997.)

urrect them. A voice from heaven will say, "Come up hither," and their enemies will see them taken to heaven by a "cloud."

"I'm not saying we are Jesus," Nettles wrote to her daughter. "It's nothing as beautiful but it is almost as big. . . . We have found out, baby, we have this mission before coming into this life. . . . It's in the Bible in Revelation." The Two taught that on six occasions God had sent souls to earth to uplift humanity: (1) Adam. (2) Enoch, who was Adam reincarnated in a new vehicle. (3) Moses. (4) Elijah. (5) Jesus. (6) Bo and Peep.

The Two's first move was to open an occult bookstore in Houston. After it failed in 1973, they took to the road to gather converts. A group was started in Los Angeles called Guinea Pig. Applewhite was Guinea, Nettles was Pig. Soon they were calling their movement HIM (Human Individual Metamorphosis). Later it became TOA (Total Overcomers Anonymous). Because they considered themselves shepherds to a flock of sheep, Applewhite took the name of Bo, and Nettles became Peep. Over the years they liked to give themselves other whimsical names such as Him and Her, Winnie and Pooh, Tweedle and Dee, Chip and Dale, Nincom and Poop, Tiddly and Wink. Eventually they settled on the musical notes Do and Ti.

In a 1972 interview in the *Houston Post,* Nettles said her astrological work was assisted by Brother Francis, a nineteenth-century monk. "He stands beside me," she said, "when I interpret the charts." Both Do and Ti constantly channeled voices from superbeings who lived on the Evolutionary Level Above Human, or Next Level (the Kingdom of Heaven).

It all sounds so childish and insane, yet those who attended early cult meetings, mostly on college campuses, have testified to the pair's persuasive rhetoric. Early converts were mainly young hippies, drifters, and New Agers, disenchanted with other cults, eager to be told what to believe and do.

In 1975 about twenty followers were recruited in the seaside village of Waldport, Oregon, then taken to eastern Colorado, where they expected to board a flying saucer and be carried to the Next Level. It was a vague region ruled by the great EGB (Energy God Being). When the spacecraft failed to appear, it was such a blow to Bo and Peep that they plunged underground for seventeen years.

There was a period when The Two preached that the "Demonstration" foretold in Revelation 11 would occur. As I said earlier, this would be their assassination, followed by their resurrection and journey to the Heavenly Kingdom in a spacecraft that the Bible called a cloud. "The chances it won't happen," Applewhite told a *New York Times* reporter in 1976, "are about as great as that a rain will wash all the red dirt out of Oklahoma." The interview got him fired as choir director of St. Mark's Episcopal Church in Houston.

Never sexual lovers, the peculiar pair surfaced in the mid-seventies as leaders of about fifty followers who wandered with them here and there. They camped out or lived in motels with funds donated by wealthy recruits or obtained from odd jobs and occasional begging. For several years they hunkered down in a camp near Laramie, Wyoming. HIM was now a full-blown cult with members strictly regulated by what they called the Process. Recruits assumed new names. Sex, alcohol, tobacco, and pot were taboo. Ti, whom Do always considered his superior, died of cancer in 1985 after losing an eye to the disease. Until his suicide, Do said he was in constant communication with Ti, who had reached the Next Level.

Precise details about the cult's nomadic history remain obscure. Do convinced his sheep that they, too, were aliens from the Next Level, now incarnated in a body they called the soul's container, vehicle, instrument, or vessel. When the time was right, they would all be teleported to one of the spaceships operated by angels.

Time reported (August 27, 1979) that cult members were then wearing hoods and gloves, obeying "thousands" of rules, studying the Bible intensely, and undergoing periods in which they communicated with each other only by writing. It's not easy to believe, but the cult received so much media attention in the late seventies that a TV series called *The Mysterious Two* was planned. A pilot actually aired in 1982 featuring John Forsythe and Pricilla Pointer as The Two.

After the Internet became widely available to the public, the cult intensified its recruiting by way of a website called Heaven's Gate. A few followers had developed sufficient skills not only to go online, but also to run a service called Higher Source that designed websites for customers.

In 1996 the cult rented a sprawling Spanish-style villa, with pool and tennis court, in Rancho Santa Fe, a few miles north of San Diego. The rent was $7,000 a month. Members began the day with prayers at 3 A.M., ate only two meals a day, had their hair cropped short, and wore baggy clothes to make themselves look genderless and unsexy. Their lives were more regulated than the lives of soldiers. Guns were stored just in case government forces attacked them as they had the Branch Davidians in Waco. Meticulous plans were drawn for a mass suicide as soon as the higher beings gave them a "marker" in the heavens. The marker, Do decided, was the giant spacecraft that psychics had convinced him (perhaps verified by the voice of Ti) was following Comet Hale-Bopp. A lunar eclipse on March 23, 1997, may have strengthened the sign.

Haunting videotapes were made on which the smiling and happy sheep said how joyfully they were looking forward to escaping from their vehicles and from a doomed planet. "We are happily prepared to leave 'this world' and go with Ti's crew," they posted on their website. Evidently they believed their beloved Ti was on the Hale-Bopp spacecraft!

As everyone now knows, eighteen men and twenty-one women put themselves to sleep with phenobarbital mixed into pudding or applesauce and washed down with vodka. Plastic bags were tied over their heads so they would suffocate in their sleep. Faces and upper bodies of the "monks," as they called themselves, were neatly covered with a square of purple cloth. All thirty-nine were dressed alike—black shirts, black pants, and black Nike running shoes. The last two to die were women with bags on their heads but unshrouded by a purple covering.

The most perplexing aspect of these ritualized deaths was the neatly packed travel bags beside their bunks, and a $5 bill and some quarters in each of their pockets. Did they expect the superbeings to take the bags along with their souls? And what use did they suppose the money would have when they reached the spaceship?

The odor of rotting "vehicles" was so strong that the first police at the scene on March 26 suspected poison gas.

To me the saddest aspect of this insane event was the firm belief, expressed on the incredible videotapes, that cult members were killing themselves of their own free will. Nothing could have been more false. Although Do always told his robots they were free to go at any time—and hundreds had done just that—so powerful was his control over the minds of those who stayed that they believed anything he said, obeyed every order. Autopsies showed that Do and seven of his followers had been surgically castrated.[2]

Do said he was dying of cancer. Yet his autopsy showed no sign of cancer or any other fatal illness. The wild-eyed expression on his face, reproduced on the covers of both *Time* and *Newsweek,* was not a look of illness. It was a look of madness.

Media reports have made fun of the belief that our bodies are mere containers and that in our next life we will be given glorious new bodies. This, of course, is exactly what St. Paul taught, and what conservative Christians, Jews, Muslims, and most Eastern faiths believe. Similar mixtures of New Testament doctrine with New Age nonsense is what makes so many recent cults appealing to converts with Christian backgrounds. Members of Heaven's Gate firmly believed that Jesus was an extraterrestrial sent to Earth like Do and Ti to collect as many souls as possible and lead them upward to acquire new containers. When Jesus finished his work, he went back to heaven in a UFO.

The great adventist movements in America—Seventh-day Adventism, the Jehovah's Witnesses, and Mormonism—are flourishing today as never before despite the long delay in Jesus' Second Coming. None of the major adventist faiths recommend suicide, but there may be more suicides by other weird little cults that surely will be capturing the minds of lonely, gullible souls.

So pervasive is the worldwide belief in alien UFOs that a London company recently offered to insure anyone against abduction, impregnation, or attack by aliens. About four thousand people, mostly in England and

2. Earlier cults have recommended castration to curb male sexual passions. The most famous intentional castration in the history of Christendom was the self-castration of Origen, the greatest of the church fathers next to Augustine. Unable to curb his lust for young women pupils, Origen sliced off his testicles. He was sorry later that he'd done it. Do may have felt a kinship with Origen, who believed in a plurality of inhabited worlds, the pre-existence of human souls, and the ultimate salvation of all sinners, including the devil.

the United States, bought policies. In October 1996, Heaven's Gate paid $1,000 for a policy covering up to fifty members for $1 million each. After their mass suicides, the London firm decided to abandon its insurance against space aliens.

Some insight into the sort of people who were followers of Bo and Peep can be gained from a sad story distributed by the Associated Press in early April 1997. Lorraine Webster, age seventy-eight, now living in Rollo, Missouri, abandoned her husband in 1978 to help found Heaven's Gate. She left the cult only because of a health problem. Her daughter was among those who died at Rancho Santa Fe.

Was Lorraine Webster disturbed by the suicides? Not in the least. Like all cult members, she doesn't like to call her cult a cult. It was a "movement." Do, she told the reporter, was a "kind and wonderful man." She misses her daughter but admires her for acting "like an angel." Ti frequently talks to Ms. Webster. She recently appeared at Webster's window in the form of a "chirping bird."

At the fifth annual Gulf Breeze (Florida) UFO Conference, March 21–23, 1997, Courtney Brown announced that his psychics' most recent remote viewing of Comet Hale-Bopp showed that the spacecraft was no longer there. It had moved, he said, to a spot behind the sun. Evidently this news failed to reach Do and his sheep. However, if Do was in touch with Ti on the ship, he probably would have taken her word over Brown's.

The *Village Voice*, covering the cult in its December 1, 1975, issue, included a prophetic passage: "The whole operation has lost that silvery crazy glitter. Now it seems black, dark, and a little ugly. It has the smell of ordinary death."

References

"UFO Cult Mystery Turns Evil." Victoria Hodgetts, in *Village Voice*, December 1, 1975, pp. 12–13.

"Looking For the Next World." James S. Phelan, in *New York Times Magazine*, February 29, 1976, pp. 12–13, 58–64.

UFO Missionaries Extraordinary. Edited by Hayden Hewes and Brad Steiger. Pocket Books, 1976.

"Flying Saucery in the Wilderness." *Time,* August 27, 1979, p. 58.

How and When Heaven's Gate May Be Entered. Published by the cult on the Internet, 1995, 200 pages.

Cover story in *Newsweek,* April 7, 1997.

Cover story in *Time,* April 7, 1997.

Cover story in *People,* April 14, 1997.

Numerous reports in *New York Times, Washington Post, Los Angeles Times,* and other newspapers during the weeks following the discovery of the suicides on March 26, 1997.

"The Faithful Among Us." Howard Chua-Eoan, in *Time,* April 14, 1997.

"De-Programming Heaven's Gate." *New Yorker,* April 14, 1997.

"Eyes on Glory: Pied Pipers of Heaven's Gate." Barry Bearak, in *New York Times,* national edition, April 28, 1997.

"Heaven Couldn't Wait." John Taylor, in *Esquire,* June 1997.

"UFO Mythology: Escape to Oblivion." Paul Kurtz, in *Skeptical Inquirer,* July/August 1997.

Heaven's Gate Cult Suicide in San Diego. New York Post staff. Harper Paperbacks, 1997.

Addendum

The mass suicides at Rancho Santa Fe were followed by several copycat suicides and attempted suicides. On March 31, 1997, Robert Nichols, fifty-eight, who lived alone in a trailer in Yuba County, California, was found dead with a plastic bag over his head and a purple shroud covering his body. A suicide note said, "I am going on the space ship with Hale-Bopp to be with those who have gone before me." Apparently he had no connection with Heaven's Gate, though he was mired in both astrology and ufology.

In Waynesville, North Carolina, on April 1, Ronald Wayne Parker and his friend Chan Patrick Alfred Naillon, both in their twenties, tried to kill themselves in hope of catching a ride on the spaceship trailing Hale-Bopp. Unable to obtain phenobarbital, they settled for rat poison, which they mixed with applesauce and chased down with vodka. They were released after a hospital stay.

On May 6, two former cult members, Wayne Cook and Chuck Humphrey, were found in a motel room four miles from Rancho Santa Fe. Both had attempted suicide. One was found dead. The other, unconscious, was taken to a hospital and placed in intensive care.

The two men wore black clothing and black sneakers and had packed tote bags. They had $5 bills and three quarters in their pockets. The number five apparently had some special significance for the cult (quarters are 5×5 cents).

Cooke and Humphrey made two videotapes that they sent here and there, and that were played on television. They invoked Do's favorite metaphor of the caterpillar who becomes a butterfly. In a similar way, they said they were not dying but merely leaving their vehicles for life on a higher plane. In a letter to CNN, Cooke wrote: "I simply cannot stay here any longer and I am leaving because it is time for me to leave. I'd rather gamble on missing the bus this time than staying on this planet and risk losing my soul."

Cooke's wife, Suzanne Sylvia, was among the thirty-nine who killed themselves on March 26 at Rancho Santa Fe.

In spite of the mass suicides, there are still sad remnants of the cult around and active. In July 1994, five true believers spoke in a classroom at the University of Illinois at Chicago. For two hours they bored an audience of some forty students with recitations of the doctrines they had been taught by Bo and Peep.

In Courtney Brown's new book *Cosmic Explorers* (1999), which is discussed in the addendum of the preceding chapter, Chapter 22 is titled "A Companion No Longer." It covers the remote viewing of the Hale-Bopp comet by advanced students at Brown's Farsight Institute in Atlanta.

Brown's students, and Brown himself, did indeed detect a mysterious "Companion" near the comet. Brown saw inside the Companion male and female humanoids, wearing uniforms and working at desks arranged in concentric curved rows. He entered the mind of one of the high-ranking men with a "deep mind probe." The man was "in a deep funk," severely troubled by worry and fear.

Apparently the Companion could disappear at will, turning on and off like a car's headlights. Brown admits that he knows nothing about the nature of the Companion or its purpose. He describes it as a "dimensionless window or portal of extraterrestrial origin, and more, much of which we could not understand. . . . It could perhaps shrink to a molecular size, or expand potentially to a size greater than that of our solar system. It is simply physically impossible to know the fixed extensions of such advanced technology." Here is how Brown ends his chapter:

> The object known as the Hale-Bopp Companion is clearly not a companion of a comet. Yet I do not know where it is currently. These data suggest that a governmental facility with a large domed structure is monitoring it, or perhaps searching for it. The humans in the facility are apparently very aware of the ET object. It is spherical, shiny, and has a golden color. The governmental facility is apparently in an area that has a harsh and cold climate, most likely in the northern or northwestern parts of the United States. Perhaps the Dakotas might fit this description.
>
> The high-ranking personnel who are in charge of monitoring this object appear to be quite worried about the implications of its existence. Some of these personnel would apparently like simply to destroy the object. There is a tension in the air as they continue to observe it or to search for it.
>
> I do not know the purpose of this object. It obviously can move through space, and it possibly can be used to transport beings and physical equipment. The object itself may be any size. I simply do not know its current size, nor can I discern its size from these data.
>
> This object has caused more controversy than any previous UFO sighting except the famous Roswell crash. The existence of this object is probably extremely important in the drama that grips humanity today. Its existence is also evidence that some ETs want to force us into a psy-

chological crisis of awareness. They do not appear to want to hide their ships or their controversial technology. They want us to see them.

Brown does not mention his appearances on the Art Bell show or the role played by his institute's remote viewing in the mass suicides of Heaven's Gate believers.

Marshall Herff Applewhite, leader of Heaven's Gate, in his videotaped farewell message to the world. (Reuters/NBC/Archive Photos)

Part VIII

More

Fringe

Science

CHAPTER 20

Thomas Edison, Paranormalist

Thomas Alva Edison (1847–1931) was the world's most famous, most prolific inventor. I will spend little time on biographical details because they are easily found in encyclopedias or in the more than sixty books about Edison. Nor will I be concerned with whether his 1,093 patents are all to be credited to his undisputed genius or to the work of many assistants. It has been said that his greatest invention was the invention factory, or research team. Many of his inventions were improvements on earlier work by others. (*Most* inventions are.) The incandescent lightbulb, for example, had a long history before Edison found better filaments. His one great, indisputably original invention was the phonograph.

This also is not the place to discuss Edison's foibles: his temper tantrums, his lust for money, his efforts to purloin ideas, his boasts about war weapons that never existed, or his disastrous relations with his two wives and his children. These are aspects of Edison's character I did not know about when forty years ago I wrote an adulatory article about him for *Children's Digest* (November 1954).

My intent here is to focus on Edison's changing religious opinions, his lifelong interest in psychic phenomena, and his gullibility. My main sources are two biographies—Robert Conot's *Thomas A. Edison: A Streak of Luck* (1979) and Wyn Wachorst's *Thomas Alva Edison: An American Myth* (1981)—and the chapter on Edison in Martin Ebon's *They Knew the Unknown* (1981).

In his youth Edison was an outspoken freethinker. He greatly admired Thomas Paine's *Age of Reason,* but unlike deist Paine, Edison did not believe in God, the soul, or an afterlife. At that time Edison was a pantheist who liked to call nature the "Supreme Intelligence," indifferent and merciless toward humanity. His friend Edward Marshall interviewed him for the *New York Times* (October 2, 1910). "There is no more reason to believe that any human brain will be immortal," Edison declared, "than there is to think that one of my phonograph cylinders will be immortal. . . . No, the brain is a piece of meat mechanism—nothing more than a wonderful meat mechanism."

Edison's words, occasioned by the death of William James, generated an uproar of opposition from Christians of all stripes. He was soundly trounced by Cardinal Gibbons. *Columbian Magazine,* a Catholic periodical, devoted an entire issue to attacking what it called "Edison's materialism."

Then something happened to Edison on the way to his laboratory. In an interview titled "Edison Working on How to Communicate with the Next World," in *American Magazine* (October 1920), B. C. Forbes—he later founded *Forbes* magazine—revealed that Edison not only had come to believe in an afterlife, but was actually working on an electrical device for communicating with the dead! (See also Austin Lescarboura's "Edison's Views on Life after Death," in *Scientific American,* October 30, 1920.)

Nothing is known about the kind of machine Edison had in mind, although it is known that he conducted experiments with it. It was probably some sort of telephone using greatly amplified electromagnetic waves.

Martin Ebon quotes the following remarks made by Edison to the *Scientific American* interviewer:

> If our personality survives, then it is strictly logical and scientific to assume that it retains memory, intellect, and other faculties and knowledge that we acquire on this earth. Therefore, if personality exists after what we call death, it's reasonable to conclude that those who leave this earth would like to communicate with those they have left here.
>
> . . . I am inclined to believe that our personality hereafter will be able to affect matter. If this reasoning be correct, then, if we can evolve an instrument so delicate as to be affected, or moved, or manipulated . . . by our personality as it survives in the next life, such an instrument, when made available, ought to record something.
>
> Certain of the methods now in use are so crude, so childish, so unscientific, that it is amazing how so many rational human beings can take any stock in them. If we ever do succeed in establishing communication with personalities which have left this present life, it certainly won't be through any of the childish contraptions which seem so silly to the scientist.

Christian leaders here and abroad welcomed Edison into their ranks as a theist who now believed in immortality. *Scientific American,* in the article cited earlier, ran a photograph of Edison pouring liquid from a flask into a beaker. The caption read: "Thomas A. Edison—the world's foremost inventor—who is now at work on an apparatus designed to place psychical research on a scientific basis."

Although Edison never became a Christian, Mina Miller, his young and pretty second wife (she was eighteen years his junior), never wavered from her devout Methodist upbringing. Conot (p. 427) calls her "an unreconstructed fundamentalist who . . . thought evolution a plot of Satan." I had the pleasure of meeting her when I was a small boy. My parents had taken me to Chautauqua, New York, where the Edisons maintained a summer cottage. I rang their doorbell to ask for the great man's autograph. He was not at home, but Mrs. Edison graciously promised to have him send it to me, which he did.

Another interview by Marshall titled "Has Man an Immortal Soul?" appeared in the *Forum*'s November 1926 issue. Edison now speaks of the "soul," and refers to God as both a "Great Power" and a "Creator." "Today the preponderance of probability greatly favors belief in the immortality of the intelligence, or soul, of man," Edison said. He praises Christianity as the wisest and most beautiful of world religions, seeing it as evolving toward a faith with less emphasis on doctrines and more on the moral code of Jesus. Theologians should stop debating creeds, Edison emphasized, and devote more time to "pile up the evidence . . . which no fool skeptic can demolish."

In later interviews that produced newspaper headlines around the world, Edison conjectured that the human mind was composed of billions of infinitesimal particles that are responsible for intelligence and memory. He thought they came from outer space, bringing wisdom from other inhabited planets. After we die, they may disperse, or they may swarm like bees and enter other human skulls, he said. Edison liked to call his particles "little people." Occasionally, he said, they get into conflict with one another. Here is how he put it in his diary:

> They fight out their differences, and then the stronger group takes charge. If the minority is willing to be disciplined and to conform there is harmony. But minorities sometimes say: "To hell with this place; let's get out of it." They refuse to do their appointed work in the man's body, he sickens and dies, and the minority gets out, as does too, of course, the majority. They are all set free to seek new experience somewhere else.

Edison was fascinated throughout his long life with the occult. In his thirties he became intrigued by the writings of that amusing mountebank Madame Helena Petrovna Blavatsky, the great guru of theosophy. Edison attended meetings in New York of the Theosophical Society and was awarded some sort of diploma. A firm believer in PK (psychokinesis), he tried to start pendulums swinging by mind control, but the results were negative. He also attempted to confirm telepathy by experiments with electric coils around the heads of human receivers and transmitters. Ebon quotes from Edison's diary: "Four among us first stayed in different rooms, joined by the electric system. . . . Afterwards we sat in the four corners of the same room, gradually bringing our chairs closer together toward the center of the room, until our knees touched, and for all of that, we observed no results."

It was Edison's good friend Henry Ford who introduced Edison to the magician Berthold Reese (1841–1926), better known as Bert Reese. He was a fat, bald-headed little man with pop eyes and a round face like a cherub. Born in what is now Poland, "Dr." Reese, as he liked to call himself, traveled widely around Europe performing what magicians call "mental magic" for celebrities and royalty. He liked to wear on his tie a huge diamond pin given to him by the king of Spain, and an even larger diamond on a finger ring. Many leading parapsychologists believed he had extraordinary psi powers.

Reese specialized in what is called "billet reading." He would ask someone to write something on a piece of paper, which he would fold and either hide or destroy. Reese would then pretend to read the message by ESP. His methods were well known to honest magicians of the time. There are scores of ways to accomplish billet reading.

Houdini was so impressed by Reese's skill that in a letter to Conan Doyle (April 3, 1920) he said that Reese "is without doubt the cleverest reader of messages that ever lived." Houdini urged Doyle to have a "séance" with Reese if he ever visited New York City, where Reese was then living, to see if "you can fathom his work."

In his book *Paper Magic,* Houdini refers to Reese in a footnote as "in my estimation, the greatest pellet reader that ever lived. (A pellet is a billet rolled into a ball.) I had a séance with Dr. Reese, and if it had not been for my many years of experience as an expert, I might have been mystified by his adroit manipulations and uncanny deductions."

Edison was the most famous person to be totally bamboozled by Reese.

Like so many scientists who tumble for psychic charlatans, Edison considered himself far too intelligent to be fooled, and of course it never occurred to him to seek explanation from a magician. When an article in the *New York Graphic* unveiled some of Reese's techniques, Edison was furious. He sent the newspaper a letter in which he said:

I am certain that Reese was neither a medium nor a fake. I saw him several times and on each occasion I wrote something on a piece of paper when Reese was not near or when he was in another room. In no single case was one of these papers handled by Reese, and some of them he never saw, yet he recited correctly the contents of each paper.

Several people in my laboratory had the same kind of experience, and there are hundreds of prominent people in New York who can testify to the same thing.

Houdini wrote to Doyle on August 8, 1920:

You may have heard a lot of stories about Dr. Bert Reese, but I spoke to Judge Rosalsky and he personally informed me that, although he did not detect Reese, he certainly did not think it was telepathy. I am positive that Reese resorts to legerdemain, makes use of a wonderful memory, and is a great character reader. He is incidentally a wonderful judge of human beings.

That he fooled Edison does not surprise me. He would have surprised me if he did not fool Edison. Edison is certainly not a criterion, when it comes to judging a shrewd adept in the art of pellet-reading.

The greatest thing Reese did, and which he openly acknowledged to me, was his test-case in Germany when he admitted they could not solve him.

I have no hesitancy in telling you that I set a snare at the séance I had with Reese, and caught him cold-blooded. He was startled when it was over, as he knew that I had bowled him over. So much so that he claimed I was the only one that had ever detected him, and in our conversation after that we spoke about other workers of what we call the pellet test—

Foster, Worthington, Baldwin et al. After my séance with him, I went home and wrote down all the details.

The letters are quoted from *Houdini and Conan Doyle: The Story of a Strange Friendship* (1932), by Bernard Ernst and Hereward Carrington. Joseph Rinn, in *Sixty Years of Psychical Research* (1950), has a good description of one of Reese's billet-reading performances, with an explanation of how he did it.

The best account of Reese's methods is "Bert Reese Secrets," by magician Ted Annemann, in the 1936 Summer Extra issue of his periodical, *The Jinx*. It includes a photograph of Reese, his hand holding a cigar that he habitually smoked during his performances because it made it easier to palm a folded billet. Annemann writes that Harvard's distinguished German-born philosopher and psychologist Hugo Münsterberg (1863–1916) "became such a believer in Reese's powers that he was preparing a book on him when death prevented its finish." I was unable to verify this. Like his friend William James, Münsterberg believed in both God and immortality, but unlike James he was a well-known skeptic of the paranormal who had a great record of exposing mediums and other psychic charlatans by carefully contrived traps.

There is evidence that Edison thought he himself had ESP. At any rate, there is no question that his powers of precognition were poor. Here are some of his failed predictions that I found in *The Experts Speak* (1984), an amusing anthology by Christopher Cerf and Victor Navasky, and elsewhere:

"The talking motion picture will not supplant the regular silent motion picture. . . . There is such a tremendous investment in pantomime pictures that it would be absurd to disturb it." (*Munsey's Magazine,* March 1913.)

"It is apparent to me that the possibilities of the aeroplane, which two or three years ago was thought to hold the solution to the [flying machine] problem, have been exhausted, and that we must turn elsewhere." (*New York World,* November 17, 1895.)

"The radio craze . . . will die out in time so far as music is concerned. But it may continue for business purposes." (Quoted by Conot in his biography of Edison, p. 424.)

"Sammy, they will never try to steal the phonograph. It is not of any commercial value." (Edison to Sam Insull, an assistant, as quoted by Conot, p. 245.)

"In fifteen years, more electricity will be sold for electric vehicles than for light." (Quoted in *Science Digest,* February 1982.)

Edison's worst prediction had to do with what was called the "war of the currents." Nikola Tesla and others believed that alternating currents were the best way to transmit high-voltage electricity over long distances. Edison stubbornly insisted that only direct current should be used. "There is no plea which will justify the use of high-tension alternating currents, either in a scientific or a commercial sense. They are employed solely to reduce investment in copper wire and real estate. . . . My personal desire would be to prohibit entirely the use of alternating currents. They are as unnecessary as they are dangerous. . . ." (I quote from David Milsted's article "Even Geniuses Make Mistakes," in *New Scientist,* August 19, 1995.)

Edison's influence on science fiction is covered in the entry "Edisonade," in the *Encyclopedia of Science Fiction* (revised edition, 1995), edited by John Clute and Peter Nichols. The literature starts with the Tom Edison, Jr., sequence of dime novels, by Edward Ellis. Edison is also portrayed as a character in a French novel, *Tomorrow's Eve* (1886), by Villiers de l'Isle-Adam, and in Garrett P. Serviss's *Edison's Conquest of Mars* (1898). For more recent references consult the *Encyclopedia of Science Fiction.*

In the introduction to his book, Conot sums up his opinion of Edison this way:

The Edison that I discovered was a lusty, crusty, hard-driving, opportunistic, and occasionally ruthless Midwesterner, whose Bunyanesque ambition for wealth was repeatedly subverted by his passion for invention. He was complex and contradictory, an ingenious electrician, chemist, and promoter, but a bumbling engineer and businessman. The stories of his inventions emerge out of the laboratory records as sagas of audacity, perspicacity, and luck bearing only a general resemblance to the legendary accounts of the past.

John Brooks, reviewing Conot's book for the *New York Times Book Review* (February 25, 1979), was even harsher:

> Thomas Alva Edison's beliefs and habits were those of a crackpot and a bum. Rats lived happy and undisturbed in his laboratory; he often slept in his clothes, because he believed that changing or taking them off induced insomnia; he thought Richard Wagner was Jewish; he was a disastrous husband and father; he all but starved himself to death because he believed that food poisons the intestines; his own company in Europe coined the cable name "Dungyard" for him.

Addendum

Lives of Edison continue to be written. Two massive biographies have recently been published: Neil Baldwin's *Edison: Inventing the Century* (1995) and Paul Israel's *Edison: A Life of Invention* (1998).

Jesse Glass wrote from Japan to question the claim made by Forbes that Edison actually built a device for communicating with the dead. He enclosed material containing a passage from a *New York Times* interview with Edison (October 15, 1926). Forbes had visited him, Edison said, "on one of the coldest days of the year. His nose was blue and his teeth were chattering. I really had nothing to tell him, but I hated to disappoint him, so I thought up this story about communicating with spirits, but it was all a joke."

I don't believe it. In 1948, Dagobert Runes's Philosophical Library published *The Diary and Sundry Observations of Thomas A. Edison.* As edited by Runes, the diary contains a section reprinting what Edison had to record in 1920 about life after death and his spirit communication device:

> I have been at work for some time building an apparatus to see if it is possible for personalities which have left this earth to communicate with us. If this is ever accomplished, it will be accomplished, not by any oc-

cult, mysterious, or weird means, such as are employed by so-called mediums, but by scientific methods. If what we call personality exists after death, and that personality is anxious to communicate with those of us who are still in the flesh on this earth, there are two or three kinds of apparatus which should make communication very easy. I am engaged in the construction of one such apparatus now, and I hope to be able to finish it before very many months pass.

If those who have left the form of life that we have on earth cannot use, cannot move, the apparatus that I am going to give them the opportunity of moving, then the chance of there being a hereafter of the kind we think about and imagine goes down.

On the other hand, it will, of course, cause a tremendous sensation if it is successful.

CHAPTER 21

What's Going On at
Temple University?

In recent years Temple University, a distinguished coeducational institution in Philadelphia, has become a center for the promulgation of some of the wildest aspects of pseudoscience. It all began in 1986 when Richard J. Fox, chairman of Temple's board of trustees, met with some fringe scientists in London. He became impressed by their difficulties in getting work published that went beyond "mainstream paradigms." "Paradigm" is still a favorite buzzword of maverick scientists and those who write about them.

There was a crying need, Fox decided, for an organization that would permit fringe scientists to interact with mainstream scientists and provide a forum for discussing their results. If Temple University would sponsor such a center it could make certain that high academic standards were maintained. Here is how Fox described the purpose of such an organization:

> The Center's overall objective is to create a legitimate place and environment where scientists, researchers, and thinkers from all areas of scientific and intellectual endeavor can come together and discuss their thoughts, projects, and ideas no matter how revolutionary, with complete confidence and comfort.

Temple's president, Peter Liacouras, agreed. The center's mission, he declared, was "to examine critically frontier research projects that hold promise of future breakthroughs."

Temple's Center for Frontier Sciences, as it is now called, was founded in 1987. Since then it has sponsored a raft of conferences, and more than fifty lectures on Temple's main campus. Its periodical *Frontier Perspectives,* issued twice a year, has grown to more than eighty pages. I had not seen a copy until physicist C. Alan Bruns, at Franklin and Marshall College, in Lancaster, sent a copy of Vol. 7, No. 1, 1998, to CSICOP's office, which in turn forwarded it to me.

Reading through its pages I could hardly believe my eyes. I had expected the magazine to be concerned with such outstanding frontiers as superstring theory, the nature of dark matter, the genetic origins of altruism, how organic molecules fold so rapidly, speculations about a "multiverse" in which endless universes, each with a unique set of laws, explode into reality, or supercomputers operating with quantum mechanics.

The "frontiers" covered in this peculiar journal are nothing of the sort. They are reports on research so far removed from reputable science that it is no wonder academic journals refuse such papers. Let me quickly review a few topics that dominate the Fall/Winter 1998 issue of this magazine.

Homeopathy is one of the center's favorite "frontiers." This is the nineteenth-century crank contention that certain substances, diluted to a degree that no molecules of the substance remain, have great potency in curing an enormous variety of ailments. Because homeopathic remedies consist of nothing but distilled water, it becomes necessary for its defenders to assume that, in some mysterious manner totally unknown to chemists, the water retains a "memory" of its vanished substances.

Cyril Smith, a British electrical engineer, writing on "Is a Living System a Macroscopic Quantum System?" relates "homeopathic potencies" to the earth's electromagnetic fields that cause dowsing rods to turn. The Center for Frontier Sciences obviously regards the ancient art of water witching as another of today's science "frontiers." In 1989 it sponsored a conference on dowsing, chaired by Terry Ross, identified as a "well-known dowser."

Nancy Kolenda, executive editor of *Frontier Perspectives,* writes, "The participants found the meeting to be a learning experience that gave them the opportunity to develop their skill as dowsers. . . ." A second conference on dowsing, titled "Bioinformation Sensing and Sensitivity to Geophysical Fields," was held later in 1989 in Germany.

Writing on "Three Frontier Areas of Science That Challenge the Paradigm" (*Frontier Perspectives,* Vol. 3, No. 1, 1992), Beverly Rubik, for seven years director of the center, conjectures that dowsing is related to ELF (extremely low frequency) electromagnetic waves. ELF waves are another major concern of the center, especially the alleged terrible effects on human health of ELF waves bombarding us from overhead electrical wires. The other two major concerns of the center, Rubik asserts, are alternative medicines and the nature of consciousness.

Glen Rein, in a paper on how quantum fields heal, finds that such fields, rather than electromagnetic fields, are what alter the properties of water and give it healing powers. Like other authors in this journal, Smith and Rein write in a mind-numbing technical jargon almost impossible to understand.

F. Fuller Royal and Gregory Olson discuss "Illness as a Delusion." They actually believe that illness has no reality! (Are they Christian Scientists? one wonders.) Illness, according to these authors, is caused by "mental delusions" in a mind that is not confined to the brain but is active in every atom of our body. Homeopathic remedies, they maintain, "are patterns of nonlinear waves that resonate with similar thought programs located in the memory field of the subconscious mind and with perturbations in the conscious mind field. These medicines are capable of eliminating delusional programs located in the memory field that serve as the foundation for illness."

Before birth, according to Royal and Olson, we existed outside of time in a region of "pure light." Time entered our lives only after we descended "into a lower earthly vibration." A developing fetus is strongly influenced by its mother's emotional state. Homeopathic drugs are "harmonic non-linear soliton waves in resonance with subconscious negative programs. . . . The energy of homeopathic medicine will collapse a negative thought program . . . making it no longer available to enter the conscious field." Delusions can also be banished by a second "treatment modality" the authors call TFT, "Thought Field Therapy."

In 1990 the Center for Frontier Sciences sponsored a conference on homeopathy in Baden-Baden, Germany. Among its speakers was Jacques Benveniste, a French homeopath whose work on "water with a memory" was so thoroughly discredited a year or two earlier.[1] Nancy Kolenda writes that the conference "ended on a high note with a unanimous decision to move forward in a global cooperation in promoting homeopathic research."

Beverly Rubik is also gung ho for homeopathy. Her paper on "Frontiers of Homeopathic Research" ran in the Vol. 2, No. 1, 1991, issue of *Frontier Perspectives*. Bruns, who tipped me off to this bizarre periodical, said in a letter that he heard Rubik lecture at a regional meeting of the American Association of Physics Teachers. Her stirring defense of psychic powers included an account of her experience with Russia's "magnetic women" who suspend metal objects on their foreheads and chests. Rubik showed slides of herself with a spoon stuck to her forehead! Bruns was amazed that no one in the audience laughed or snickered.

Enough about homeopathy. To regard its revival today as a frontier science is comparable to calling a revival of phrenology or palmistry a frontier science. It has been said that anyone today who believes in phrenology

---

[1]After INSERM, France's medical research agency, closed down Benveniste's laboratory, he opened his own Digital Biological Laboratory south of Paris. He recently claimed to have transmitted "water memory" over the Internet, using e-mail. And he is suing two Nobel Prize winners, physicist Georges Charpak and biologist François Jacob, and physicist Claude Hennion for writing unkind things about him. On Benveniste's monumentally flawed homeopathic research, see Chapter 4 of my *On the Wild Side* (Prometheus, 1992).

ought to have his or her head examined. The same can be said of today's homeopathy enthusiasts who are unable to distinguish cures from placebo effects.

Here are a few other fields of parascience presented favorably in the Fall/Winter 1998 issue of *Frontier Perspectives:*

In "Is Dead Matter Aware of Its Environment?" Peter Graneau argues that all particles of matter are aware of all other particles regardless of how far away they are. He thinks Newton's physics is superior to Einstein's and likens the blindness of establishment scientists today to the blindness of those Italian professors who refused to accept Galileo's experiment of dropping two different weights from the Tower of Pisa. Graneau is unaware that this experiment was never performed.

Dan Kenner, an acupuncturist, defends the thousands of herbal remedies sold in Oriental shops. He doesn't mention the shops in India where the herbs are quite different from those in China and Japan. Kenner introduces a word that was new to me—"nosology." It is not a study of noses, but the science of classifying diseases. Homeopathy, Kenner tells us, is an example of "empirical nosology"—that is, a way of classifying illnesses based on careful research.

Roger Taylor favorably reviews a self-published book—he calls it a "gem of science"—titled *Waves in Dark Matter.* The author, O. Ed Wagner, has done experiments which show that these waves, previously undetected, are responsible for what he calls the hitherto "unexplained ability" of trees to raise water up their trunks. A Chinese biophysicist, Taylor adds, has done work which suggests that these elusive W-waves play a role in the spacing of acupuncture points on human bodies. "Without doubt a new and important chapter has been opened in the science of life," Taylor concludes. Another book under review extols the great benefits of green tea in inhibiting cancer, dental caries, and other ailments.

The magazine's funniest paper is "On the Nature of Tarot," by Inna Semetsky, identified as someone at Columbia University's Teachers College. Semetsky defends the validity of Tarot card readings. The practitioner uses the random arrangement of the shuffled cards to tune in to fields that Semetsky relates to David Bohm's "implicate order," Jung's archetypes and concept of synchronicity, Heisenberg's uncertainty principle, and karma. In addition to space's three dimensions, and the dimension of time, there is a fifth dimension consisting of consciousness. Because time is a "parameter" of this fifth field, it allows Tarot readers to tap into Jung's "collective unconscious," part of the fifth field, and learn about future events. Semetsky urges the introduction of Tarot into mental health professions.

Prominent believers in ESP, PK, and precognition (some no longer living) who have been active in the center's conferences and/or contributors to its journal include Brian Josephson, Rupert Sheldrake, Andrija Puharich (author of a book about Uri Geller), Robert Jahn and his psychic assistant Brenda Dunne, Glenn Olds, Willis Harmon, Helmut Schmidt, Ramakrishna Rao, Harold Puthoff, Stephen Braude, David Griffin, Fred Wolfe, and many others.

Another piece of evidence that Temple University is sliding into absurdity involves UFOs. On Temple's faculty as an associate professor of history is David Jacobs, one of our nation's most energetic promoters of the reality of human abductions by extraterrestrial aliens. His first book, *The UFO Controversy in America* (Indiana University Press, 1975), is an expanded version of his doctoral thesis at the University of Wisconsin. *Secret Life* (Simon & Schuster, 1992), his second UFO book, is devoted to first-hand accounts of abductions. *The Threat,* his latest book (also from Simon & Schuster), was published early this year. "Ph.D" appears after Jacobs's name on the jacket, and also after his name at the top of every left-hand page, a sure giveaway to the man's ego.

Although Jacobs has had no training in psychology, psychiatry, or hypnotherapy, he uses hypnotism to induce his patients (now more than seven hundred) to develop strong memories of horrendous abductions even though many patients had no such memories until hypnotized. Jacobs is convinced that five million Americans have been kidnapped at least once by aliens. One female patient, who worked in retail sales, had, according to Jacobs, a hundred abductions in one year, an average of one every three days! How did she manage to keep her job, *New York Times* reviewer Joe Queenan wanted to know.

Jacobs's patients routinely report incredible sexual molestations. The aliens extract sperm from men, eggs from women, then use them to produce a race of hybrids intended soon to take over Earth. Jacobs writes that he "desperately wishes" this not to be true, but now he "fears for the future" of his children. Jacobs isn't sure where the aliens come from, but he thinks it might be from a distant planet. They communicate with each other and with humans by telepathy. You might suppose Jacobs would look favorably on other UFO researchers using hypnosis to revive memories of abductions. Not so. For example, he regards John Mack, Harvard's embarrassing psychiatrist who has also written a book about UFO abductions, as incompetent and gullible. As for Philip Klass, the nation's top debunker of the UFO mania, Jacobs refuses even to speak to him.

The hybrids who walk among us are fiendishly clever in concealing themselves. They look and dress exactly like us. To further confuse us, the aliens plant false memories in abductee heads so that when they are returned to Earth, police think they are nuts because they talk about seeing Jesus, the Virgin Mary, Abraham Lincoln, and other notables. These fake memories are created by a technique called "mindscan," a term Jacobs coined. It never occurs to him that he himself is using a form of mindscan on his patients.

"If *The Threat* has any shortcoming," wrote Joe Queenan in his *New York Times* review (January 10, 1998), "it is its failure to explain why aliens always seem to abduct people no one ever heard of. . . . Nor does he [Jacobs] inform the reader why these seemingly omnipotent creatures have never nabbed him. Perhaps Mr. Jacobs has in fact been abducted, brainwashed, and tricked into writing this book—with the specific purpose of making a reputable publisher appear inane and UFO hunters seem even more laughable."

Aside from Klass's books and newsletters, the most powerful, most hilarious recent debunking of the UFO scene is an article by Frederick Crews in *The New York Review of Books* (June 25, 1998). Titled "The Mindsnatchers," it reviews three UFO books, one of which is Jacobs's *The Threat*. As Crews recognizes, Jacobs, like John Mack and others, is blissfully unaware of how easily false memories can be fabricated. Fortunately, these memories are less harmful than false memories of sexual abuse by human adults. Innocent fathers, mothers, and teachers have already wasted years in prison, some even under life sentences, solely on the basis of fabricated memories of sexual abuse dramatically recounted in court by children and grown-ups who were brainwashed by fanatical therapists.

Crews quotes the following passage from Jacobs's account of the memories of a patient he calls "Beverly":

> Then the hybrids told Beverly that they could take her body whenever they wanted and that she was always vulnerable and never safe. One hybrid raped her, and she was forced to perform fellatio upon another. They pinched her, twisted her skin, and hurt her without leaving marks. They pushed an unlit candle into her vagina. They then told her she had caused her children to be abducted. . . . On another occasion hybrids made her envision her six-year-old daughter walking into a room ringed with naked hybrids who had erections; she was led to believe that her daughter would be raped by all of them.

That a seemingly sane history professor could believe such obvious confabulations and keep his job at a major university is surely bewildering. Temple even allows Jacobs to regularly teach a course on ufology. There is one thing that can be said in praise of *Frontier Perspectives*. It has not yet published a paper by Jacobs, or, as far as I know, reviewed any of his preposterous books.

Addendum

My column about Temple University generated more letters than any other column reprinted in this book. I will refer briefly to the letters that were published in *Skeptical Inquirer,* followed by my replies.

Geologist Mark Wilson called attention to Temple's promotion of extreme Afrocentrism. Professor Molefi Kete Asante, chairman of the university's Department of Africology, claims that the Egyptians were blacks, and that the Greeks stole from them their culture. Wilson called this bogus history, and believes that sound knowledge is being sacrificed to a "misguided racial consciousness" designed to increase the self-esteem of black Americans.

Nancy Kolenda, director of Temple's Center for Frontier Sciences, insisted that I failed to understand the center's mission. She listed a number of distinguished scientists serving on the center's advisory and editorial boards. Inna Semetsky complained that I failed to give any reasons for denying the efficacy of Tarot card readings.

The distinguished biologist Lynn Margulis came to my defense by describing a bad experience she once had with the center. Physician Gary Posner reinforced my dim view of David Jacobs's UFO claims. Dan Kenner defended his article. He was especially incensed by my attack on homeopathy. Peter Graneau similarly defended his paper that bashed relativity theory.

I replied to these letters as follows:

It would be helpful if Temple University's Center for Frontier Sciences had a director who was a scientist. Director Nancy Kolenda, an accountant, fails completely to comprehend the distinction between reputable frontier science and crank science. To feature articles defending homeopathy, dowsing, Tarot card reading, and the notion that physical ills are mental delusions is no different from publishing articles defending palmistry, phrenology, numerology, and Velikovsky's crazy cosmology.

It is a great embarrassment to a distinguished university and its scientists that the chairman of Temple's board of trustees would squander $100,000 a year on a center and its periodical that contribute absolutely nothing to the progress of science.

Inna Semetsky is irked because I called her defense of Tarot card reading "funny," without giving my reasons for rejecting such fortune-telling. I would no more waste space trying to "refute" Tarot card reading than I would waste space trying to refute palmistry or tea-leaf reading. Ms.

Semetsky's letter, with all its references to modern physics, is even funnier than her article. That a magazine claiming to discuss matters on the "frontiers" of scientific research would publish her paper is, however, not funny, but sad.

It's not easy to have rational dialogue with someone who believes, as does Dan Kenner, in the efficacy of homeopathic drugs and dowsing. I couldn't find "phytotherapy" in my Webster's College Dictionary, but I assume it means plant therapy. Kenner's use of this term is a fine example of obfuscatory jargon. To dismiss as "nonsense" the many recent books on the genetic origin of altruism (I recommend *Unto Others: The Evolution and Psychology of Unselfish Behavior,* by Elliott Sober and David Sloan, published by Harvard University Press) is another indication of Kenner's bizarre beliefs. As for "shops," I meant the tens of thousands of shops in Japan, China, and India that sell thousands of herbal remedies (they differ from country to country) for every conceivable ailment including cataracts and venereal diseases.

Peter Graneau belongs to a tiny group of maverick physicists who think that relativity, now confirmed to the hilt, is deeply flawed. He is convinced it is refuted by every electrical motor in the world! *Newtonian Dynamics,* by Peter and Neal Graneau, is a vigorous attack on Einstein. *Newtonian Electrodynamics,* also by the Graneaus, is a similar effort to apply Newtonian theory to electrical phenomena. The title of a third book by the Graneaus, *Newton Versus Einstein,* says it all. In the paper I criticized, Peter Graneau writes, "It should be clear that nature has spoken, and as far as her remarks go, recommends the Newtonian world views."

I am pleased to learn that elsewhere Graneau recognized that Galileo's Tower of Pisa experiment is a myth. Why then does he write in his article that the "blindness" of those who defend relativity theory is "synonymous" with the blindness "displayed by the professors of the University of Pisa when they saw Galileo's cannonball and the light musket ball fall with the same velocity from the leaning tower"? Did Graneau forget what he said earlier?

Beware of the scientists who liken themselves to Galileo and their antagonists to the professors of Pisa.

Faye Flam, formerly with *Science* and now a staff writer for the *Philadelphia Inquirer,* featured my column in a lengthy article titled "Call it Freedom? Or Beyond the Fringe?" (August 31, 1998). The subhead: "Activity at Temple's Center for Frontier Sciences is the target of a columnist." "The Center is a surprisingly small operation," Flam writes, "consisting of director Nancy Kolenda and one graduate student." A photograph of Kolenda is included.

Temple's faculty members are reluctant to talk about the center, Flam says, but John Allen Paulos, a mathematician well known for his popular books, was not reluctant. He compared the center to "the scandal of the lobotomized cousin up in the attic." He told Flam that the articles in *Frontier Perspectives* were so dumb, and their arguments so pseudointellectual, as to be embarrassing to him as a Temple professor. He referred to one article as nonsensical, term-dropping "quantum drivel."

Prominent Temple scientists can't distance themselves enough from the center, Flam adds. Edward Gawlinski, head of Temple's physics department, commented: "We don't have any interactions with them." By "them" he meant those associated with the center.

My column was also the topic of Emily Nussbaum's article "Out There" in *Lingua Franca* (December/January 1999). She reports that Kolenda "has no scientific background," but claims to have a degree in accounting from Villanova University. However, the registrar's office at Villanova told Nussbaum that Kolenda took only three night-school classes and did not complete a degree.

Kolenda was naturally outraged by my column and the publicity it generated. In addition to her short letter published in *Skeptical Inquirer,* you'll find her longer response in *Frontier Perspectives* (Fall 1998). I'm bashed for basing my column on just one issue of the magazine she edits. Glancing through the issue in which her reply appeared, I found it just as bad as if not worse than the issue I focused on. Kolenda promises that the entire controversy will be aired in a future issue.

I don't know if the wealthy Richard Fox, chairman of Temple's board of trustees, has any special interest in the work of Wilhelm Reich, who invented orgone therapy and a way to use the mythical orgone energy to produce rain. I do know that his Fox Foundation, which has a $5 million endowment, gave $14,000 to James DeMeo for work with his cloudbuster. DeMeo is the nation's top Reichian researcher. On his use of long pipes to shoot orgone energy into clouds to produce rain, see Chapter 3, "Wilhelm Reich the Rainmaker," in my *On the Wild Side* (Prometheus Books, 1992). In his journal *Pulse of the Planet,* DeMeo thanks Fox for funding his cloudbusting in Israel.

Beverly Rubik is a friend of DeMeo's. When she was a graduate student at the University of California, at Berkeley, she did research on Reich's experiments in creating microcopic life forms he called bions. Rubik is said to have fabricated a segmented worm, a centimeter long, from nonliving matter. I have no idea what she is up to since she left Temple.

Peter Sturrock's Society for Scientific Information had its eighteenth annual convention in Albuquerque, New Mexico, June 3–5, 1999. Physicist Dave Thomas reported on the gathering in *Skeptical Inquirer* (September/October 1999). The papers presented were even crazier than those published in the issue of the center's journal that I summarized in this chapter.

Topics included attacks on relativity and evolution, favorable lectures on the reality of spoon bending, the miraculous creation of objects, the healing effects of prayer, alien abductions, the face on Mars, measuring bacteria contamination at distances of thousands of miles, psychokinesis, reincarnation, spirit mediums, and what Thomas described as "almost every type of paranormal . . . phenomena imaginable."

Lee Pulos, of the University of Columbia, reported on the miraculous powers of one Thomas Greene Morton, a magician turned psychic who lives in Brazil. One of his tricks was to transform an American silver dollar into a medallion made of an entirely different metal. When Thomas asked Pulos why he didn't take Morton to Randi to collect Randi's $1 million prize, Pulos snorted at the suggestion by saying that Morton had no interest in being tested.

David Jacobs, of Temple, was on hand, along with John Mack, of Harvard, to talk about alien abductions. Jacobs thinks the aliens are evil. Mack thinks they love us. Frenchman Jacques Benveniste defended his latest claim that the powers of homeopathic drugs can be trasmitted by wireless.

Leaders of the Society for Scientific Information and their selected "scientists" haven't the foggiest notion of how to distinguish authentic speculative science (e.g., superstring theory) from bogus science totally without merit.

Part IX

Religion

CHAPTER 22

Isaac Newton, Alchemist and Fundamentalist

And from my pillow, looking forth by light
Of moon or favouring stars, I could behold
The antechapel where the statue stood
Of Newton with his prism and silent face,
The marble index of a mind for ever
Voyaging through strange seas of Thought, alone.
—*William Wordsworth*, Prelude,
Book 3, lines 58–63

There are three Sir Isaac Newtons. For several centuries the best-known Newton has been the great mathematical physicist who in his early twenties invented calculus, discovered the binomial theorem, introduced polar coordinates, proved that white light was a mixture of colors, explained the rainbow, built the first reflecting telescope, and showed that the force causing apples to fall is the same as the force that guides the planets, moons, and comets and produces tides. His discoveries revolutionized physics. His genius is undisputed.

Unknown to most people, even today, are two other Newtons. One is the alchemist who struggled for decades to turn base metals into gold. The other is Newton the Protestant fundamentalist.

Newton worked quietly alone, almost in secret, on his great discoveries. His classic *Philosophiae naturalis principia mathematica* was not published until twenty years after his youthful achievements, and then only at the insistence of astronomer Edmund Halley, for whom the comet is named, and who funded the book's publication. For a large part of his life Newton's time and energy were devoted to fruitless alchemy experiments and efforts to interpret Biblical prophecy. His handwritten manuscripts on those topics far exceed his writings about physics. They constitute several million words now scattered in the rare book rooms of libraries and in private collections. The American philosopher Richard Popkin is currently working on a twelve-volume edition of these manuscripts.

Although other scientists of the time, notably Robert Boyle, were interested in alchemy, none was as obsessively occupied with such research as Newton. He read all the old books on alchemy he could find, accumulating more than 150 for his library. He built furnaces for endless experiments and left about a million words on the topic. He thought of himself as working within a secret occult tradition of wisdom that traced back to Earth's earliest history. He even suspected that the ancients had known the inverse-square law of gravity!

It is with Newton's fundamentalism that this chapter is mainly concerned. He was a devout Anglican, firmly believing the Bible to be God's revelation, although he granted that the original texts had been heavily corrupted by an unscrupulous Roman Church. He accepted the Genesis account of creation in six literal days, the temptation and fall of Adam and Eve, Noah's Ark and the universal flood, the blood atonement of Jesus, his birth by a virgin, his bodily resurrection, and the eternal life of our souls in heaven or hell. He never doubted the reality of angels and demons, and a Satan destined on judgment day to be cast into a lake of fire. Bishop James Ussher, the seventeenth-century Irish scholar, had settled on 4004 B.C. as the year of creation. Newton revised it in the wrong direction by making it five hundred years later!

Newton's universe was a vast machine operating by laws created and upheld by a personal yet transcendent deity. Infinite space was God's "Sensoriam"—the means by which he observed and controlled the cosmos.

Although for Kant and later admirers of Newton the universe was deterministic, never deviating from unalterable laws, Newton was convinced that from time to time God needed to adjust the orbits of planets to keep them free of perturbations caused by comets and other forces.

This notion that God has to tinker with the universe to repair it struck Newton's German rival, the great philosopher-mathematician Leibniz, as blasphemous. If God is perfect, omnipotent, and omniscient as Newton believed, why, Leibniz wanted to know, would he create a universe so flawed that it would require perpetual adjustments?

Newton had no use for pantheism. His God was the God of the Bible, in whose image we were created, but so wholly other that we cannot comprehend how we resemble him. Newton's greatest departure from the prevailing religion of England was his rejection of the Trinity. He was an Arian (Arianism was a forerunner of Unitarianism) for whom Jesus was indeed the divine Son of God, but in no way equal to God. Trinitarianism, Newton believed, was a crude heresy concocted by the Roman Church in the fourth and fifth centuries. He kept this belief to himself, knowing well that if it became known he would be expelled from his Cambridge college, ironically called Trinity, where he was a professor of mathematics for twenty-six years. It later would have endangered his job at the Royal Mint, where he worked for the last half of a long life. He was a diligent servant, overseeing England's currency and merciless in sending counterfeiters to the gallows. He was the first to recommend gold as a monetary standard.

For Newton, the beautiful patterns of the material universe were overwhelming evidence of God's creative powers. As an example, he singled out the fact that all the planets revolve on the same plane, in the same direction, with just enough centrifugal force to keep them from crashing into the sun. Newton was puzzled by the fact that gravity seemed to operate instantaneously at a distance. He admitted he could do no more than describe it without comprehending how it worked. Not until Einstein's general theory of relativity was gravity changed from a "force" to the movement of matter along the shortest paths in a curved space-time. As physicist John Wheeler likes to say, the stars tell space-time how to bend, and space-time tells the stars how to go.

Both space and time were for Newton absolute. Space was a fixed, infinite, unmoving metric against which absolute motions could be measured. His proof of this was the centrifugal force produced by rotating bodies, which he correctly guessed had caused the earth to bulge at its equator. It would be foolish, he argued, to suppose that a rotating bucket of water, which tossed water over its sides, was at rest with the entire universe spinning around it. There was no way, of course, he could have conceived of general relativity in which even rotary motion is relative, but how he would relish it were he to return to earth today! "Newton, forgive me," Einstein once wrote.

Nor can we fault Newton for not understanding cosmological and biological evolution. Like so many theists before and after, he saw the intricate patterns of life forms as another proof of God's wondrous handiwork. He was particularly impressed by bilateral symmetry:

> Can it be by accident that all birds, beasts and men have their right side and left side alike shaped (except in their bowels), and just two eyes and no more on either side the face, and just two ears on either side the head, and nose with two holes and no more between the eyes, and one mouth under the nose, and either two forelegs or two wings or two arms on the shoulders, and two legs on the hips, one on either side and no more?

Would Newton have accepted evolution had he lived after Darwin? If so, he would have considered it God's method of creation, though it would surely have demolished his belief in the accuracy of Genesis.

I also suspect that Newton, reincarnated today, would embrace quantum mechanics. Indeed, he thought that light consisted of particles, independent of space though somehow influenced by space. In quantum mechanics, light is both a wave and a particle called the photon.

Newton's passion for alchemy was exceeded only by his passion for Biblical prophecy. Incredible amounts of intellectual energy were spent trying to interpret the prophecies of Daniel in the Old Testament and the Book of Revelation in the New. He left more than a million words on these subjects, seeing himself as one who for the first time was correctly judging both books. Having been so successful in solving some of the riddles of God's universe, he turned his talents toward trying to answer riddles posed by God's Holy Word.

Newton was firmly convinced that Daniel and the Apocalypse, when correctly deciphered, showed that Earth's history was about to end with the Second Coming of Jesus, followed by his judgment of the quick and the dead. In his youth Newton speculated on 1867 as a possible date for the Second Coming, but in later years decided it was folly to use the Bible for predicting the future. The best we can do is recognize successful predictions *after* the events occur. Like millions of Protestants in the seventeenth century, he believed the papacy was the Antichrist foretold in the Apocalypse—an incarnation of Satan in his last, futile effort to thwart God's plan for cleansing the universe of sin. He accepted the prophecy that in the last days the Jews would return to Jerusalem and become Christians. The coming of Christ would be followed by a Millennium during which the Lord would rule over earth "with an iron hand." In his old age Newton moved the date of the Second Coming to some time after the end of the twenty-first century.

Six years after Newton's death his *Observations upon the Prophecies of Daniel and the Apocalypse of St. John* was published in London. It was reprinted in 1922, but amazingly has been unavailable since. The best summary of its contents known to me is a chapter in the second volume of Leroy Edwin Froom's *The Prophetic Faith of Our Fathers* (Review and Herald, 1950–54), a massive four-volume work by a Seventh-day Adventist historian. Froom greatly admired Newton's religious opinions. Many are shared by Adventists, including the identification of the papacy with the Antichrist and a belief that God created the universe through Jesus. Like the Adventists, Newton took the four parts of the metallic image in Chapter 2 of Daniel to symbolize the successive world powers of Babylon, Persia, Greece, and Rome. Like the Adventists, he took the growth of the "little horn" on the fourth beast in Daniel to represent the growth of the papacy. What about 666, the mysterious number of the Beast in Revelation? Like today's Adventists, Newton believed we do not yet know its meaning. It would be interesting to compare Newton's exegesis of Biblical prophecy with the Adventist classic *Thoughts on the Book of Daniel and the Revelation,* an 1882 work by Uriah Smith.

To support his conviction that the Old Testament is accurate history, Newton worked out an elaborate chronology of Earth's history, drawing on astronomical data such as eclipses and star motions and legends such as that of Jason and the Argonauts, which he took to be genuine events. With incredible ingenuity he tried to harmonize Biblical history with secular histories of the ancient world. It is sad to envision the discoveries in mathematics and physics Newton might have made if his great intellect had not been diverted by such bizarre speculations.

Newton's writings on Biblical prophecy are so huge an embarrassment to his admirers that to this day they are downplayed or ignored. The long essay on Newton in the *Encyclopaedia Britannica*'s famous eleventh edition devotes only one brief paragraph to Newton's Bible studies. They are not mentioned at all in the fourteenth edition, and are allowed one paragraph in the *Macropaedia* of the current fifteenth edition.

What did Newton think of his great discoveries in physics? Amazingly, he seems to have considered them little more than youthful recreations. In a memorable, often quoted, passage he likened himself to a little boy "playing on the seashore, and diverting myself in now and then finding a smoother pebble or a prettier shell than ordinary whilst the great ocean of truth lay all undiscovered before me."

Newton's peculiar, introverted, self-absorbed personality is still an enigma. Contemporaries noticed his melancholy countenance. Although he occasionally smiled, he almost never laughed. A lifelong bachelor, he had not the slightest interest in sex. A few Freudian analysts, stressing the death of his father before he was born, have suggested that Newton was a repressed homosexual. The main evidence is that in his middle age Newton became infatuated with Nicolas Fatio de Duillier, an eccentric Swiss disciple twenty years his junior. Gale Christianson, in his 1984 biography of Newton, *In the Presence of the Creator,* strongly doubts any sexual activity between the pair took place, but adds: "On the other hand, their correspondence—with its lavish praise, requited loneliness at separation, and melancholy swings of mood—bears haunting overtones of an ill-fated romance. The final break itself appears to have been prefigured in their agonizing desire to share the same chambers, a desire quite possibly overridden by the fear of what might happen if they were to attempt it."

Newton had no interest in music or art, and once dismissed poetry as "ingenious fiddle-faddle." He never exercised, had no recreational hobbies and no interest in games, and was so preoccupied with his work that he frequently forgot to eat or would eat standing up to save time. He had few friends, and even to them he was often quarrelsome and vindictive. In one of his letters to John Locke, his best friend among British philosophers, he wrote:

Being of opinion that you endeavoured to embroil me with woemen & by other means I was so much affected with it as that when one told me you were sickly & would not live I answered twere better if you were dead. I desire you to forgive me this uncharitableness.

Locke wrote back to grant forgiveness, and to express undimmed love and esteem.

Newton seldom credited other scientists with earlier work that had influenced him. Always insistent on getting full credit for his discoveries, he bitterly accused Leibniz, whose metaphysics he despised, of stealing his invention of calculus. It is now known that the two discoveries were independent. Newton's was earlier, but Leibniz had a superior notation.

A few years after the publication of *Principia,* Newton suffered a massive mental breakdown that took a year or more to overcome. It was marked by severe insomnia, deep depression, amnesia, loss of mental ability, and paranoid delusions of persecution. In recent years a few scholars have suggested he may have suffered from mercurial and other toxic metal poisoning caused by his alchemical experiments. Others have conjectured that throughout his life he was a manic-depressive with alternating moods of melancholy and happy activity. His breakdown was only the worst of such episodes.

When Newton's manuscripts on alchemy were sold in 1936 at a Sotheby auction, the economist John Maynard Keynes was the major buyer. In a brilliant speech on Newton, given at the Royal Society's Newton Tercentenary Celebration in 1947, Keynes spoke of having gone through some million of Newton's words on alchemy and found them "wholly devoid of scientific value." Newton's "deepest instincts were occult, esoteric—with a profound shrinking from the world—a rapt, consecrated, solitary perusing his studies by intense introspection, with a mental endurance perhaps never equaled."

As for Newton's discoveries in mathematics and physics, Keynes believed they resulted less from experiments than from an incredible intuition. Later Newton would dress them up with formal demonstrations and proofs which had little to do with the insights that seemed to enter his head by sheer magic. Keynes put it this way:

> In the eighteenth century and since, Newton came to be thought of as the first and greatest of the modern age of scientists, a rationalist, one who taught us to think on the lines of cold and untinctured reason. I do not see him in this light. I do not think that anyone who has pored over the contents of that box which he packed up when he finally left Cambridge in 1696 and which, thought partly dispersed, have come down to us, can see him like that. Newton was not the first of the age of reason. He was the last of the magicians, the last of the Babylonians and Sumerians, the last great mind which looked out on the visible and intellectual world with the same eyes as those who began to build our intellectual inheritance rather less than 10,000 years ago. Isaac Newton, a posthumous child born with no father on Christmas Day, 1642, was the last wonderchild to whom the Magi could do sincere and appropriate homage.

My major references for working on this essay are Richard Westfall's great biography of Newton, *Never at Rest* (1980); Frank Manuel's *The Religion of Isaac Newton* (1983); Bernard Cohen's sixty-page article on Newton in the *Dictionary of Scientific Biography* (1974); and two valuable papers by Richard Popkin: "Newton and the Origins of Fundamentalism," reprinted in *The Scientific Enterprise* (1992), edited by Edna Ullmann-Margalit; and "Newton and Fundamentalism," in *Essays on the Context, Nature, and In-*

fluence of Isaac Newton's Theology (1990), by James Force and Popkin. Popkin makes a strong case for the enormous influence of Newton's Biblical exegesis on the early history of Protestant fundamentalism.

Addendum

On March 15, 1996, at three-thirty, after I typed the last page of the column that became the basis of this chapter, an astonishing coincidence occurred. I glanced out my study's window. It had just stopped raining. A drop of water, on a leaf outside the window, sparkled red. I moved my head slowly from side to side and saw the drop run through all the colors of the rainbow. The sun happened to be at just the precise spot in the sky in relation to my eyes. Of course I thought at once of Newton. I felt as if he had spoken to me from the Great Beyond.

Byron, another English poet who admired Newton, in Canto 10, Stanza 1, of *Don Juan* described Newton's discovery of gravity:

> *When Newton saw an apple fall, he found*
> *In that slight startle from his contemplation—*
> *'Tis said (for I'll not answer above ground*
> *For any sage's creed or calculation)—*
> *A mode of proving that the earth turn'd round*
> *In a most natural whirl, called "gravitation";*
> *And this is the sole mortal who could grapple,*
> *Since Adam, with a fall, or with an apple.*

Richard Westfall, who taught history and philosophy of science at Indiana University, and who spent twenty years researching his definitive biography of Newton, died in 1997. The following year saw the publication of yet another biography, *Isaac Newton: The Last Sorcerer,* by Michael White. A third biography, *Newton: The Making of Genius,* by Patricia Fara, was published in 2002.

William Blake, *Newton* (1795) (Tate Gallery, London/Art Resource, New York)

CHAPTER 23

Farrakhan, Cabala, Baha'i, and 19

The two greatest sacred books of world religions are the Bible and the Koran. Given the belief that God directed the writing of each, it is not hard to understand how the faithful could imagine that their divinely inspired text would contain hidden mathematical structures proving the book's supernatural origin.

With respect to the Old Testament, this kind of numerology reached its zenith among the Cabalists—Jewish mystics whose roots trace back to pre-Christian times but who flourished mainly in medieval and Renaissance centuries. They believed that letters of the Torah (the Old Testament's first five books) concealed thousands of secret meanings. These mysteries were extracted from the text by a process called gematria. A number was assigned

to each Hebrew letter. These numbers were then added, multiplied, and manipulated in other ways to disclose subtle correlations and hidden meanings.

Similar techniques were widespread among Christian mystics. They assigned numbers to both Hebrew and Greek letters of the Bible. In recent
years the most indefatigable Christian gematrist was the Russian mathematician Ivan Nikolayevich Panin, who died in Canada in 1942. His vo-
luminous writings furnished what he believed was irrefutable proof that the
Bible was written by the Almighty.

The most preposterous book of Christian gematria published in the
United States this century was *Theomatics: God's Best-Kept Secret Revealed*
(Stein & Day, 1972). It was written by Jerry Lucas, an All-American bas-
ketball star with the New York Knicks, and his fundamentalist friend Del
Washburn. According to the dust jacket, their book "scientifically proves
that a Mind, far beyond human capabilities . . . planned, constructed, and
formed every word in the Bible." Their technique is an original form of
absurd gematria. The naive authors seem totally unaware of the vast earlier literature along similar lines. (For details about this book, see my review in *Order and Surprise,* Prometheus Books, 1983.)[1]

Today's Christian fundamentalists have lost interest in gematria, but the
practice still thrives among extremely Orthodox Hebrew scholars. The latest development in this dismal field springs from the use of computers to
analyze the Torah. In the 1980s, three Israeli mathematicians, Doron Witz-
tum, Eliyahu Rips, and Yoav Rosenberg, made an extensive search through
the Torah, using a sophisticated computer program, for secret words and
phrases. The words emerged when the computer checked every nth letter,
where n can take any value. They were amazed to find the names of dozens
of famous rabbis of later centuries. Future events were predicted. For ex-

---

1. I suspect that Jerry Lucas is now much ashamed of this book, but not Del Washburn. In 1994, Scarborough House published his *Theomatics II: God's Best-Kept Secret Revealed.* Steve Abbott, reviewing this ridiculous book in the *Mathematical Gazette* (July
1996), wrote: "I have wondered how to dispose of it. I have rejected throwing it in the
dustbin, as I can't bear to destroy books. I could give it to a charitable shop, but it might
be bought by someone impressionable, and I wouldn't want to be responsible for the consequences. So it will stay on my shelf unread, unless a *Gazette* reader asks me for it."

ample, in one sequence of letters they found the Hebrew words for *Sadat, president, shot, gunfire, murder,* and *parade.* They took this to be a prediction of Anwar Sadat's assassination in 1981. Even Norman Schwarzkopf's name turned up!

Media around the world gave all this widespread coverage in 1995 when they reported on an article by psychiatrist Jeffrey Satinover in the October issue of *Bible Review.* Titled "Divine Authorship?" the article discussed two papers by the Israeli mathematicians on what they call ELS (equidistant letter sequences) in Genesis.

As a simple experiment, I considered only the first fifteen words in Lincoln's Gettysburg Address and checked every nth letter for n equal to 2 through 10—that is, letters separated by one through nine letters. I found thirty-two three-letter words and the following four-letter words: *sort, soar, Nero, huts, hoot,* and *NATO.* Imagine how many longer words would turn up in books as long as Genesis or a play by Shakespeare, and allowing n to vary from 2 through 100. In ancient Hebrew there are no vowels. This results in considerable vagueness over what word is intended. I could have found much longer words in the first fifteen words of Lincoln's speech had I been allowed to insert vowels between consonants.

In their highly technical 1994 paper "Equidistant Letter Sequences in the Book of Genesis" (*Statistical Science,* Vol. 9, pp. 429–38), the Israeli mathematicians claim that they applied their program to a Hebrew translation of Tolstoy's *War and Peace.* Although they found plenty of words, they insist that nothing turned up comparable to the words and phrases they found in Genesis. Why Jehovah would go to such trouble to hide words in the Torah beats me. It strikes me as blasphemous to turn God into a whimsical dabbler in crude wordplay.

I know of no Catholics or liberal Protestants who take ELS numerology seriously, but American fundamentalists have latched onto it as proof that God wrote the Old Testament. On a television show of June 6, 1996, I heard Hal Lindsey defend with great enthusiasm the Israeli findings. Lindsey is the Protestant fundamentalist who has written a series of best-sellers about the soon emergence of the Antichrist and the coming of Jesus. Other fundamentalists around the world are now trumpeting the Israeli research.

It would be interesting to know what fundamentalist mathematicians would come up with if they applied an ELS computer program to the Greek letters of the New Testament.

Wall Street Journal reporter Calmetta Coleman, in an article titled "Seminar Tries Science to Revive Faith" (*Wall Street Journal,* November 11, 1996), reports that ELS numerology is now being heavily promoted by a Jerusalem-based organization called Aish HaTorah, founded by Rabbi Noah Weinberg. The organization runs a movement called Discovery that conducts hundreds of seminars in Jewish synagogues and schools around the United States—seminars designed to win secular Jews back to Orthodox Judaism.

Readers may be surprised to learn that over the centuries Islamic fundamentalists have squandered equally incredible amounts of time and energy on numerology intended to prove that Allah, through the angel Gabriel, dictated the Koran to Muhammad. Americans got a brief glimpse of this number play when Louis Farrakhan spoke for two and a half hours to blacks who assembled in Washington, D.C., for the October 16, 1995, "Million Man March." Here is an excerpt from Farrakhan's speech:

> There in the middle of this mall is the Washington Monument, 555 feet high. But if we put a 1 in front of that 555 feet, we get 1555, the year that our first fathers landed on the shores of Jamestown, Virginia, as slaves.[2]
>
> In the background is the Jefferson and Lincoln Memorial. Each one of these monuments is 19 feet high. Abraham Lincoln, the 16th president, Thomas Jefferson, the 3rd president, and 16 and 3 make 19 again. What is so deep about this number 19? Why are we standing on the Capitol steps today? That number 19, when you have a 9, you have a womb that is pregnant, and when you have a 1 standing by the 9, it means that there's something secret that has to be unfolded. . . .

2. Farrakhan's history is flawed. The first slave ship did not arrive on our shores until 1619.

In *The Million Man March Home Study Guide Manual,* published by Farrakhan before the march, he included a map of the Capitol's Mall, on which lines connecting major buildings revealed a concealed pyramid. Minister Farrakhan is convinced that Washington, D.C., was carefully designed by Masons to symbolize the Great Pyramid of Egypt, the Sphinx, the Valley of Kings, and so on. He approves this because he believes that Masons were followers of Muhammad. Do not Shriners wear fezzes and mention Allah in their sacred rites? Farrakhan's *Home Study Guide* calls attention to the number of U.S. presidents who were Masons, or what he calls "Moslem Sons." (See Charles Freund's article "From Satan to Sphinx: The Masonic Mysteries of D.C.'s Map," *Washington Post,* November 5, 1995.)

On October 15, 1996, on Ted Koppel's *Nightline,* Farrakhan said that the earth has long been observed by higher beings in UFOs and that our government is concealing this from the public. An enormous flying saucer he called "Ezekiel's wheel" may soon appear in our skies. UFO believers have long regarded the first chapter of Ezekiel as an ancient description of a gigantic flying saucer.

The Nation of Islam was founded in the 1930s by Wallace D. Fard. He was succeeded by the late Elijah Muhammad. Malcolm Little (Malcolm X), the movement's most energetic leader, was assassinated by members of the movement. The Nation of Islam split in the late 1970s, with most of the faithful going to Elijah's son Wareth Deen, and a small group taking Farrakhan (born in 1933 in the Bronx as Louis Eugene Walcott) as their leader.

I am not sure where Farrakhan got his fascination with 19; probably from Elijah and his followers. In the *Final Call,* a paranoid newspaper published by the Nation of Islam, Tynetta, one of Elijah's former wives, writes a column titled "Unveiling the Number 19." It was Dr. Rashad Khalifa, whom we will meet in the next chapter, who introduced her to the wonders of 19 in the Koran. In addition to wild numerology, Tynetta is deep into Egyptian pyramidology and ufology. In Isaiah 19:19 she found a reference to an Egyptian "altar to the Lord," which she says is "about 19 feet high."

In a September 27, 1995, column, Tynetta told how Farrakhan, ten years earlier, was taken by a beam of light from the top of a Mexican pyramid to a small alien spacecraft. It then docked with a mother ship fifteen hundred times larger. There he heard the voice of Elijah Muhammad tell him that Ronald Reagan was planning to bomb Libya!

For many centuries Islamic mystics, especially the Sufis, regarded 19 as sacred, and in recent times the number has played a central role among the Baha'is. The Baha'i offshoot of Islam began in 1844. That was the year a young Persian merchant who took the name of Bab—the word signifies a gate to faith—declared himself to be, like John the Baptist, the herald of a great new "manifestation" of Allah. Because Islam does not recognize a prophet since Muhammad, Babism was bitterly denounced as a heresy. The Bab was imprisoned by the Persian government and, in 1850, was executed by a firing squad. Thousands of Bab'is were also murdered.

In 1863 another Persian, taking the name of Baha'u'llah (Glory of God), declared himself to be the very manifestation foretold by the Bab. His and the Bab's writings constitute the Baha'i scriptures. After long periods in prison, Baha'u'llah was allowed to spend his final days near Haifa, now part of Israel. After he died in 1892, his eldest son took over the movement called Baha'i, meaning "the followers of Baha'u'llah."

The son's trip to the United States in 1912 so energized the nation's small group of Baha'is that the religion began to grow. There are now said to be more than five million Baha'is around the world. The largest numbers are in India (some two million), with more than 130,000 in the United States and about 300,000 in Iran (formerly Persia). A third of the U.S. Baha'is are African-Americans. The movement is now growing most rapidly in Africa and India. You'd never guess what native state harbors the most Baha'is. It is not California, but South Carolina!

The main appeal of the Baha'i faith is its teaching that all religions contain basic truth, although the faithful believe that the truth reached its purest form in the teachings of Baha'u'llah. After death, the soul leaves the body for endless progress in another realm of being, about which we can know nothing except that it is as different from our world as our world differs from our mother's womb.

Baha'i also stresses the equality of the sexes, the equality of races, and the harmony of religion and science. It sees humanity as slowly evolving toward a world-state free of wars and injustices and speaking a world language. Among prominent Americans who became Baha'is are the singer and motion picture star Vic Damone, the artist Mark Toby, who died in 1976, and the late jazz trumpet player Dizzy Gillespie.

The Baha'i world headquarters are in Haifa, and its U.S. headquarters are in Wilmette, a northern suburb of Chicago. Wilmette is the site of a magnificent Baha'i temple. Six similar houses of worship have been built in other cities around the world. The religion has no clergy or rituals, although members are urged to pray daily, to fast nineteen days during Ala, the last month of the Baha'i year, and to make at least one pilgrimage to Haifa. Women are exempt from the fast period if they repeat "Glorified be God, the Lord of Splendour and Beauty" $5 \times 19 = 95$ times on each of the nineteen days.

Since their origin, the Baha'is have endured violent persecution by Islamic fanatics. This reached a terrible climax in the early 1980s when Shiite leaders under Khomeini tried to destroy Baha'ism in Iran. More than two hundred Baha'is were killed, hundreds more imprisoned. Thousands lost their homes and possessions. Mobs desecrated Baha'i halls, sacked their temples, cemeteries, and shrines. Baha'i schools and corporations were taken over. The House of Bab, a holy shrine, was demolished. The violence subsided after 1985, but Baha'i religious activities in Iran remain forbidden.

Baha'is are entranced by 9 and 19, both numbers having roots in Islamic tradition. Nine stands for the nine manifestations of the transcendent, wholly other, unknowable Allah. They are Moses, Buddha, Zoroaster, Confucius, Jesus, Muhammad, Krishna, the Bab, and Baha'u'llah. The Wilmette temple has nine sides, nine doors, nine pillars, nine arches, nine ribs in the dome, and nine fountains. At each entrance are $2 \times 9 = 18$ steps. Baha'i temples elsewhere have similar structures. A nine-pointed star is the most widely used symbol of Baha'i faith. Baha'is like to point out that 9, being the highest digit, is a symbol of perfection, and that Baha'i has a numerical value of 9 according to a system of numerology common in the Bab's day.

Islam's other sacred number, 19, is mentioned in the twenty-fifth verse of the Koran's Sura 74, titled "The Hidden Secret." Anyone who denies that the Koran came from Allah, this chapter says, will suffer forever in hell under the supervision of nineteen angels. "What mystery doth God intend for this number?" another verse asks. The mystery and sacredness of 19 was constantly stressed by early Islamic mystics, from whom it passed into the teachings of the Bab, down to today's Baha'is and to Minister Farrakhan.

The Bab took great pride in having $2 \times 9 = 18$ chief disciples who, together with himself, made 19. Babism included a gematria that rivaled the Cabala in its obsession with hidden meanings based on assigning numbers to the twenty-eight Arabic letters. The number 19 was found everywhere, notably as the sum of the numerical values of the Arabic and Persian letters of *Wahid,* the word for *One,* one of Allah's chief names.

The product of 19 and 19 is 361, which the Babists called "the number of all things." It is the number of days in the Babist calendar, still used by the Baha'is. Their year consists of nineteen months, each with nineteen days. The extra four days are intercalary, with a fifth day added during leap years. When the Bab made his pilgrimage to Mecca, he sacrificed nineteen lambs.

Wherever possible, the Babists divided things into nineteen parts. The years form nineteen cycles, of which we are now in the eighth. The Bab's followers even tried to base a coinage on nineteen, but had to abandon it as impractical. On the first day of each month, the Baha'is, following the Bab's instructions, assemble for the Feast of the Nineteenth Day. The movement is governed by the NSA (National Spiritual Assembly), which consists of nine members elected annually by a fixed number of $171 = 9 \times 19$ deligates from local assemblies in the continental United States.

Farrakhan's initials are L. F. Using the cipher $A = 1$, $B = 2$, and so on, his initials add to 18, one short of 19. Maybe he should change his first name to Moses or Muhammad.

The next chapter will tell the tragic story of Dr. Rashad Khalifa, who tried to convince the world that the prevalence of 19 in the Koran proved it was written by Allah, and who was stabbed to death in 1990 for his heretical opinions.

Addendum

According to *Final Call,* the Nation of Islam's newspaper, the Mother Wheel is a "human-built planet" now in orbit around the earth. It measures half a mile by half a mile, and is carrying 150 planes armed with bombs designed for the destruction of Earth. Farrakhan claims to have visited it on September 17, 1985, and there spoken with Elijah Muhammad. Some members of the Nation of Islam believed that in 1999 the Mother Wheel's bombs would destroy Earth's white government and put blacks back in the power they once had in ancient Egypt.

In *Free Inquiry* (Summer 1999), Norm Allen reported that Farrakhan has informed his followers that Elijah Muhammad did not die a natural death. He was beamed aboard Ezekiel's wheel or what Farrakhan likes to call the Mother Wheel. Farrakhan challenged skeptics to dig up Muhammad's coffin to find it empty!

In 1987, Simon & Schuster made the best-seller list with *The Bible Code,* a dreadful work intended to appeal to Protestant fundamentalists and Orthodox Jews. Although the author, journalist Michael Drosnin, said he was an atheist and did not believe the Old Testament was the inspired word of God, he was convinced that the hidden messages found by the Israeli scholars were far too miraculous to be concidences. He suggested to interviewers that maybe they were put there by some kind of alien intelligence! In a letter to the *Los Angeles Times Book Review* (August 24, 1997), he likened the discovery of a secret Bible code to Galileo's discovery that the earth goes around the sun!

On July 18, 1997, the television show *Crossfire* was devoted to Drosnin's book. John Sununu was skeptical, but Geraldine Ferraro defended the book. She said she believed in God, was fascinated by the Bible code, and couldn't believe the hidden words were coincidences. Her remarks probably sold thousands of copies of the book. It had already had an even heftier promotion by Oprah Winfrey, who likes any book that explores the paranormal.

Del Washburn, coauthor of *Theomatics,* is still arguing that *his* numerology, not Drosnin's, proves that the Bible was written by God. In 1998 he published *The Original Code in the Bible: Using Science and Mathematics to Reveal God's Fingerprints.* Washburn thinks Drosnin's code is nonsense, and that his own code is the real thing.

I find controversy over the Bible code to be so far-out, boring, and unworthy even of attack that I will content myself only with giving a few easily checked references in which the code is thoroughly discredited:

"Seek and Ye Shall Find." Sharon Begley, in *Newsweek,* June 9, 1997.

"Deciphering God's Plan." David Van Biema, in *Time,* June 9, 1997.

"He Who Mines Data May Strike Fool's Gold." Peter Coy, in *Business Week,* June 16, 1997.

"Harum-Scarum." Michael Shermer, in *Los Angeles Times Book Review,* July 20, 1997.

"Hidden Messages and the Bible Code." David E. Thomas, in *Skeptical Inquirer,* November/December 1997.

"Bible Codes, Marilyn, and El Niño." Mark Schilling, in *Math Horizons,* February 1998.

"Bible Code Developments." David E. Thomas, in *Skeptical Inquirer,* March/April 1998. See also the letters section.

"God Only Knows." Hal Cohen, in *Lingua Franca,* July/August 1998.

"Tolstoy Predicts Bulls' Sixth Championship (in Code of Course)." David E. Thomas, in *Skeptical Inquirer,* November/December 1998.

The Bible Code: Fact or Fake? Phil Stanton. Crossway, 1998.

"What Are the Chances of That?!?" Richard Morin, in *Washington Post,* April 4, 1999. The author reports on finding "M. L. King" in *Moby-Dick* near the phrase "to be killed by them," "Kennedy" close to "shoot," "Lincoln" near "killed," and "Princess Di" near "mortal in these jaws of death."

"Solving the Bible Code Puzzle." Brendan McKay, Dror Bar-Natan, Maya Bar-Hillel, and Gil Kalai, in *Statistical Science*, Vol. 14, 1999, pp. 149–73.

CHAPTER 24

The Numerology of Dr. Khalifa

In recent years the Muslim who worked the hardest in searching for instances of 19 in the Koran was the late Dr. Rashad Khalifa. After graduating from Ain Shams University in Cairo, where he ranked first in his class, he obtained a master's degree from the University of Arizona and in 1964 a doctorate in plant biochemistry from the University of California, Riverside. After two years with the Egyptian government, he became a research assistant at the University of Arizona, followed by work for the Monsanto Company in St. Louis. In 1963 he married Stephanie Hoefle, a native of Phoenix, Arizona.

During 1975 and 1976, Khalifa was science adviser to the Libyan government. He was employed by the United Nations' Industrial Development

Organization, in Vienna, before he became a senior chemist in Arizona's State Office of Chemistry in 1980. He published more than twenty scientific papers, a book titled *The Computer Speaks: God's Message to the World,* a 538-page new translation of the Koran, and numerous religious articles and pamphlets. His books and related publications are available from the International Community of Submitters (ICS), P.O. Box 43476, Tucson, AZ 85733.

In 1972, Dr. Khalifa privately printed a monograph titled *Number 19: A Numerical Miracle in the Koran.* He believed this booklet offered for the first time in history physical proof of God's existence. How? By showing beyond all doubt that the Koran must have been written by Allah, and that the text, unlike the Christian Bible, had been perfectly preserved. In a 1980 letter to me he said his results were "so overwhelming that they will inevitably shake the world," although Satan would do all he could to block the great revelation.

I have space for only a tiny selection of claims made by Dr. Khalifa. The Koran has $6 \times 19 = 114$ suras (chapters). The number of verses is $19 \times 334 = 6,346$, and the digits of that number add to 19. The Koran contains $19 \times 17,324 = 329,156$ letters. The invocation verse, called the *Basmala* ("In the name of Allah, most gracious, most merciful"), heads every sura except the ninth, but appears an extra time in the middle of Sura 27 (3×9). This invocation contains 19 Arabic letters. Its first word, *Bism,* occurs in the Koran 19 times. Its second word, *Allah,* appears $19 \times 142 = 2,698$ times. The third word, *Al-Rahman,* is repeated $19 \times 3 = 57$ times, and the fourth word, *Al-Rahmeen,* occurs $19 \times 6 = 114$ times. The sum of the verse numbers that mention Allah is $19 \times 6,217 = 118,123$. Thirty different numbers are mentioned in the Koran. These thirty numbers add to $19 \times 8,534 = 162,146$.

Each of twenty-nine suras is prefixed by a mysterious set of either one, two, three, four, or five disconnected, seemingly meaningless letters. The significance of these letters has long been a mystery. Khalifa claimed to have decoded them, and that they, too, bristle with 19s. Dozens of other findings involving 19 derive from applications of the ancient Arabic gematria.

Dr. Khalifa's research used a computer program that counted the number of times each letter occurred in each sura of the Koran. On the traditional belief that Muhammad could neither read nor write, the embedding of 19 throughout the Koran is proof, Khalifa maintained, that the Koran was dictated by Allah.

Khalifa's writings on 19 sold widely throughout Islamic countries, where they became strongly controversial, not only because his numerology tended to support the Baha'i heresy with its emphasis on 19 (see the previous chapter), but also because Khalifa rejected the last two verses of Sura 9 as spurious. Why? Because in nine places they violated the secret 19-code. So much for his earlier claim that the text of the Koran had been miraculously preserved from corruption!

Attacks on Khalifa increased in Muslim nations. His writings were banned. Muslim scholars likened his numerology to Jewish and Christian gematria. But it was much more than that. Khalifa rejected *Hadith* (words attributed to Muhammad), *Sunnah* (practices said to have been initiated by Muhammad), and *Ijma* (the consensus of scholars about Muslim doctrines). In brief, he rejected Islamic tradition. Like his counterparts among Protestant fundamentalists with respect to the Bible, he held that the Koran was the only reliable source of Islamic beliefs. Islamic leaders were incensed by his demand that the last two verses of Sura 9 be removed. Their fury increased when Khalifa's egotism grew to the point of his declaring himself the divine messenger of Allah foretold in the Koran (3:81). Death threats against him steadily mounted.

The Muslim Digest, published in South Africa, strongly attacked Khalifa as a sinister heretic. Its July/October 1986 issue disputed his word counts. The word *Allah,* it is claimed, is in the Koran 2,811 times, not 2,698 as Khalifa said. *Al-Rahman* is there 169 times, not 57 as Khalifa insisted. And so on. "Obviously," an editor wrote, "Dr. Rashad Khalifa's computer needs to brush up on its arithmetic." Earlier attacks on Khalifa are in issues July/August 1981 and March/April 1982.

In 1984, Khalifa initiated an unsuccessful lawsuit of $38 million against the National Academy of Sciences for publishing *Science and Creationism,* a booklet assuming that evolution was a godless process. On the contrary, Khalifa argued, we know from the Koran and from laws of probability that life could not have arisen without the creative actions of Allah. Each species was independently created, with evolution operating only within a species. We can be certain that God created the first humans from mud because the Koran says so in 32:7 and 15:28.

In Tucson, Dr. Khalifa founded a *masjid* (mosque), where he served as the *imam* (minister). On January 31, 1990—note that the digits of 1990 add to 19—he was assassinated, dying in the mosque's kitchen from multiple stab wounds. He was fifty-four. The alleged assassins were Black Muslims from a fanatical fundamentalist sect called Fuqra. Four members of the sect were arrested in 1993 at their compound near Buena Vista, Colorado, and later charged with murder. One man, John Williams, was convicted. The others jumped bail and vanished. The sect had earlier been implicated in the firebombing of Hare Krishna temples in Denver and Pennsylvania, and a 1983 bombing of a hotel in Portland, Oregon, owned by followers of the East Indian guru Bhagwan Shree Rajneesh.

What is one to make of Khalifa's numerology? It is, of course, no surprise that many 19s would show up in a book as long as the Koran, but Khalifa's 19s exceed the bounds of chance. The most plausible explanation is that he deceived himself by unconscious fudging. The best account known to me of how easily he could do this is in *Running Like Zebras*, a 1995 book edited by Edip Yuksel, the nation's top Khalifite.

The book contains $19 \times 6 = 114$ pages (coincidence?) that reprint a debate on the Internet between Yuksel and Daniel (Abdulrahman) Lomax, a Muslim skeptical of Khalifa's findings. He accuses the chemist of careless computer searching, of rejecting two verses of the Koran as spurious because they don't fit his calculations, and of not revealing that versions of the Koran differ in their number of words and letters, and in how they divide suras into verses. Above all, he accuses Khalifa of failing to make clear what he considers a "word."

Many Arabic words have multiple forms, and Khalifa is inconsistent in his counting rules. Sometimes he includes plural forms, sometimes not. Should a word with an affixed pronoun be called one word or two? In English the meaning of *word* is fairly clear because of spaces between words, but in Arabic there are no spaces. Even in English there is vagueness. Lomax's example is *truck*. In counting *truck* in a book should you include *trucks, trucked*, and *trucking?* One looks in vain for Khalifa's definition of *word*.

Lomax likens the doctor to those astronomers who once fancied they could see canals on Mars. He concludes: "Dr. Khalifa's claims, at best, fall into the category of pious fraud. . . . Had God intended the Qur'an to carry a code verifying its perfect preservation, he could have done it much more effectively and simply than the complex, arbitrary, and inconclusive 'code' claimed by Dr. Khalifa."

Yuksel, of course, believes he has completely demolished all of Lomax's objections. His curious book is available from his Monotheist Productions, P.O. Box 43476, Tucson, AZ 85733.

Now for some 19 number juggling, supplied in part by correspondent Monte Zerger. Nineteen is, of course, a prime. It is equal to $10^2 - 9^2$; to $1^2 + 3^2 + 3^2$; and to $3^3 - 2^3$. The number 1,729, or 19×91, was involved in a famous incident between the British mathematician G. H. Hardy and his friend Ramanujan, the Indian number-theory genius. Having taken a taxi to visit Ramanujan in a hospital, Hardy remarked that the taxicab number, 1,729, was a dull number. Ramanujan immediately replied, "No, it is an interesting number. It is the smallest number expressible as the sum of two cubes in two different ways [$12^3 + 1^3$ and $10^3 + 9^3$]." Note that the digits of 1,729 add to 19.

In 1989 (a multiple of 9) it was proved that every integer is the sum of no more than nineteen fourth powers. The smallest number requiring nineteen such powers is 79, the sum of four fourth powers of 2 and fifteen fourth powers of 1. The repeating decimal of 1/19 is the $2 \times 9 = 18$-digit number 052631578947368421. Multiply it by any number from 2 through 18 and the product has the same eighteen digits in the same cyclic order. Multiplying by 19 produces a row of $2 \times 9 = 18$ nines.

The Constitutional amendment giving women the vote was the nineteenth. The nineteenth hole in golf is the bar where golfers sink drinks like they sink puts in eighteen holes. Every nineteen years all phases of the moon fall on the same days of the week throughout the year. The Psalms is the

Bible's nineteenth book. Psalm 19 opens with "The heavens declare the glory of God, and the firmament showeth his handiwork."

The numbers most often encountered in the Bible are 12 and 7. They are reflected in our calendar, with its seven days to a week and twelve months to a year. The sum of the two numbers is 19.

Figure 1 reproduces a thing of strange beauty. The nineteen cells hold integers 1 through 19. Every straight row of cells adds to 38, or twice 19. It would make a wondrous amulet for the Baha'is.

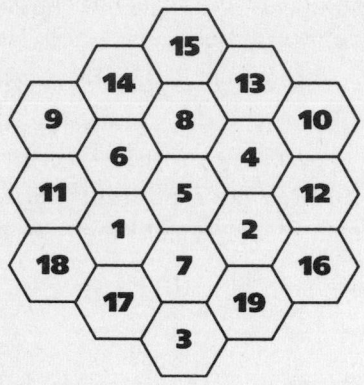

Figure 1. The only possible magic hexagon. Every straight row of cells adds to 38, or twice 19.

Addendum

Ibn Warraq is the author of a courageous book that I strongly recommend: *Why I Am Not a Muslim* (Prometheus, 1995). Warraq sent me a fascinating paper by Franz Rosenthal, Professor Emeritus of Near Eastern Languages and Civilizations at Yale University. Titled "Nineteen," it first appeared in *Analecta Biblica* (Vol. 12, 1959, pp. 304–18) and is reprinted in Rosenthal's *Muslim Intellectual and Social History,* a collection of essays published by Variorum in 1990. The essay gives a detailed history of the role of 19 in Islamic culture.

For a short time Dr. Khalifa's findings were welcomed enthusiastically in the Islamic world. The welcome quickly subsided after he began making heretical statements. One of his strongest critics was Abu Ameena Bilai Phillips, a Jamaican convert to Islam. His computer analysis of Khalifa's numerology is reported in his book *The Qur'an's Numerical: Hoax and Heresy* (1987).

Phillips accepts some of Khalifa's findings, but rejects most of them on the grounds that Khalifa fudged his statistics, mainly by failing to make clear what he considers a "word."

It's like taking the word "nevertheless," Phillips writes, to be either a single word or three different words depending on how you want a word count to come out. The final blow to Khalifa came when Sheikh 'Abdullah ibn 'Abdul Aziz ibn Baz, one of Saudi Arabia's top scholars, issued a *fatwa* branding Khalifa an apostate.

On these and other matters relating to Khalifa's numerology, see *Islamic Mysticism: A Secular Perspective* (Prometheus Books, 2000), by a British scholar who writes under the pseudonym Ibn Al Rawandi.

CHAPTER 25

The Religious Views of
Stephen Jay Gould and Darwin

*R*ocks of Ages is the clever title of the latest book by Stephen Jay Gould, Harvard's famous paleontologist and best-selling author. One of the title's two "rocks" is religion. The other is science, typified by the fossil-rich rocks that support the fact of evolution.

Professor Gould strongly opposes the notion that science and religion are irreconcilable, a claim defended in two classic works: *The Conflict Between Religion and Science* (1877) by the American scientist John William Draper, and the two-volume *History of the Warfare of Science and Theology* (1894) by Cornell historian and first president Andrew Dickson White. Both books, which Gould discusses at length, regard science and religion, especially Roman Catholicism, as locked in eternal combat.

Although Gould calls himself an agnostic inclined toward atheism, his book is a passionate plea for tolerance between the two realms. Science and religion, he contends, are examples of a principle he calls NOMA, or Non-Overlapping Magisteria. There is indeed a conflict between the two if religion is taken in the narrow sense of a creed that requires God's miraculous interventions in history and refuses to accept the overwhelming evidence for evolution. Such superstitions, by entangling the two magisteria, generate mutual enmity. If, however, religion is taken in a broader sense, either as a philosophical theism free of superstitions or as a secular humanism grounded on ethical norms, then Gould sees no conflict between the two magisteria. Not that they can be unified in a single conceptual scheme, but they can flourish side by side like two independent nations at peace with one another.

Science, Gould reminds us, is a search for the facts and laws of nature. Religion is a spiritual quest for ultimate meaning and for moral values that science is powerless to provide. To echo Kant and Hume, science tells us what *is,* not what *ought* to be. "To cite the usual clichés," Gould writes, "we get the age of the rocks, and religion retains the rock of ages; we study how the heavens go, and they determine how to go to heaven." There is no mention of John Dewey, but Gould's theme is not far from the essence of Dewey's little book *A Common Faith*.

Gould quotes liberally from the letters of Charles Robert Darwin, who, together with Thomas Henry Huxley, is one of his two greatest heroes. These quotations sent me to *The Life and Letters of Charles Darwin* (1887) by his botanist son Francis. One chapter deals entirely with Darwin's slow disenchantment with Christianity and his eventual decision to call himself an agnostic. The term had been coined by Huxley, known in his day as "Darwin's Bulldog" for his vigorous defense of natural selection and his unremitting attacks on the crude Protestant fundamentalism of England's prime minister William Ewart Gladstone.

As a youth Darwin firmly believed the Bible to be the inspired word of God. His Anglican father wanted him to become a clergyman, and Charles actually spent three years at Cambridge preparing for ordination. Although he gradually lost his faith, he always remained tolerant and respectful of the views of his Christian friends and associates, especially of the devout beliefs of his wife.

Darwin married his cousin Emma Wedgwood, who bore him ten children. They loved each other deeply, but throughout their otherwise happy marriage each agonized over their irreconcilable religious differences. Janet Browne, in her splendid biography *Charles Darwin* (1995), reprints one of Emma's letters to Charles, written before they married, in which she implores him to give up his habit of "believing nothing until it is proved." Darwin called it a "beautiful letter," and wrote on its envelope, "When I am dead, know how many times I have kissed and cried over this."

The death of their daughter Anne intensified Darwin's antipathy toward Christianity and widened the religious rift between Emma and himself. She never abandoned her faith. As a widow she may have died still tormented (as Browne puts it) by the thought that "she might not meet him [Charles] in heaven." Some biographers have even speculated, though without evidence, that Darwin's chronic illnesses were the psychosomatic consequences of the theological divide between him and his beloved wife.

Darwin's religious tolerance is at the heart of Gould's book. He even praises Pope John Paul for his 1996 statement that evolution is no longer just a theory, but a well-established fact that Catholics should accept provided they insist that immortal souls were infused into the evolved bodies of the first humans. Gould sees this as a major step on the part of Rome's magisterium toward accepting the NOMA principle.

"Nature is amoral," Gould writes, "not immoral. . . . [It] existed for eons before we arrived, didn't know we were coming, and doesn't give a damn about us. . . . Nature betrays no statistical preference for being either warm and fuzzy, or ugly and disgusting. Nature just is—in all her complexity and diversity, in all her sublime indifference to our desires. Therefore we cannot use nature for our moral instruction, or for answering any question within the magisterium of religion."

Although science is powerless to furnish ethical rules or proofs of God, neither is it capable of ruling out the possibility of a deity or the existence of moral imperatives based on a common human nature. Here is how Gould accurately summarizes Darwin's acceptance of NOMA:

Darwin did not use evolution to promote atheism, or to maintain that no concept of God could ever be squared with the structure of nature. Rather, he argued that nature's factuality, as read within the magisterium of science, could not resolve, or even specify, the existence or character of God, the ultimate meaning of life, the proper foundations of morality, or any other question within the different magisterium of religion. If many Western thinkers had once invoked a blinkered and indefensible concept of divinity to declare the impossibility of evolution, Darwin would not make the same arrogant mistake in the opposite direction, and claim that the fact of evolution implies the nonexistence of God.

Let's turn now to how Darwin himself, in carefully chosen words, expressed his religious opinions, with great humility and honesty, in correspondence quoted by his son. Here is a paragraph from a letter of 1860:

One word more on "designed laws" and "undesigned results." I see a bird which I want for food, take my gun and kill it, I do this *designedly*. An innocent and good man stands under a tree and is killed by a flash of lightning. Do you believe (and I really should like to hear) that God *designedly* killed this man? Many or most persons do believe this; I can't and don't. If you believe so, do you believe that when a swallow snaps up a gnat that God designed that that particular swallow should snap up that particular gnat at that particular instant? I believe that the man and the gnat are in the same predicament. If the death of neither man nor gnat are designed, I see no good reason to believe that their *first* birth or production should be necessarily designed.

In another 1860 letter, written to botanist Asa Gray, Darwin had this to say:

With respect to the theological view of the question. This is always painful to me. I am bewildered. I had no intention to write atheistically. But I own that I cannot see as plainly as others do, and as I should wish to do, evidence of design and beneficence on all sides of us. There seems to me too much misery in the world. I cannot persuade myself that a

beneficent and omnipotent God would have designedly created the Ich-
neumonidae with the express intention of their feeding within the living bodies of Caterpillars, or that a cat should play with mice. Not
believing this, I see no necessity in the belief that the eye was expressly
designed. On the other hand, I cannot anyhow be contented to view this
wonderful universe, and especially the nature of man, and to conclude
that everything is the result of brute force. I am inclined to look at everything as resulting from designed laws, with the details, whether good or
bad, left to the working out of what we may call chance. Not that this
notion *at all* satisfies me. I feel most deeply that the whole subject is too
profound for the human intellect. A dog might as well speculate on the
mind of Newton. Let each man hope and believe what he can. Certainly
I agree with you that my views are not at all necessarily atheistical. The
lightning kills a man, whether a good one or bad one, owing to the ex-
cessively complex action of natural laws. A child (who may turn out an
idiot) is born by the action of even more complex laws, and I can see no
reason why a man, or other animal, may not have been aboriginally produced by other laws, and that all these laws may have been expressly designed by an omniscient Creator, who foresaw every future event and
consequence. But the more I think the more bewildered I become; as indeed I probably have shown by this letter.

From a letter of 1873:

What my own views may be is a question of no consequence to any one
but myself. But, as you ask, I may state that my judgment often fluctuates. . . . In my most extreme fluctuations I have never been an Atheist
in the sense of denying the existence of a God. I think that generally (and
more and more as I grow older), but not always, that an Agnostic would
be the more correct description of my state of mind.

From a letter of 1879:

It is impossible to answer your question briefly; and I am not sure that
I could do so, even if I wrote at some length. But I may say that the im-

possibility of conceiving that this grand and wondrous universe, with our conscious selves, arose through chance, seems to me the chief argument for the existence of God; but whether this is an argument of real value, I have never been able to decide. I am aware that if we admit a first cause, the mind still craves to know whence it came, and how it arose. Nor can I overlook the difficulty from the immense amount of suffering through the world. I am, also, induced to defer to a certain extent to the judgment of the many able men who have fully believed in God; but here again I see how poor an argument this is. The safest conclusion seems to me that the whole subject is beyond the scope of man's intellect.

In 1876, Darwin wrote a candid autobiography intended to be read only by his wife and children.[1] Francis, in the biography of his father, gives a series of excerpts from this autobiography in which Darwin writes about his religious opinions. I quote this section in its entirety:

> Whilst on board the *Beagle* I was quite orthodox, and I remember being heartily laughed at by several of the officers (though themselves orthodox) for quoting the Bible as an unanswerable authority on some point of morality. I suppose it was the novelty of the argument that amused them. But I had gradually come by this time, i.e., 1836 to 1839, to see that the Old Testament was no more to be trusted than the sacred books of the Hindoos. The question then continually rose before my mind and would not be banished—is it credible that if God were now to make a revelation to the Hindoos, he would permit it to be connected with the belief in Vishnu, Siva, etc., as Christianity is connected with the Old Testament? This appeared to me utterly incredible.
>
> By further reflecting that the clearest evidence would be requisite to make any sane man believe in the miracles by which Christianity is supported,—and that the more we know of the fixed laws of nature the more incredible do miracles become,—that the men at that time were igno-

1. An unexpurgated edition of Darwin's autobiography, edited by his granddaughter Nora Barlow, was published in 1958 and is currently available as a Norton paperback. Earlier editions of the autobiography had been heavily censored by Darwin's family, mainly to remove Darwin's biting criticisms of some of his contemporaries.

rant and credulous to a degree almost incomprehensible by us,—that the Gospels cannot be proved to have been written simultaneously with the events,—that they differ in many important details, far too important, as it seemed to me, to be admitted as the usual inaccuracies of eye-witnesses;—by such reflections as these, which I give not as having the least novelty or value, but as they influenced me, I gradually came to disbelieve in Christianity as a divine revelation. The fact that many false religions have spread over large portions of the earth like wild-fire had some weight with me.

But I was very unwilling to give up my belief; I feel sure of this, for I can well remember often and often inventing day-dreams of old letters between distinguished Romans, and manuscripts being discovered at Pompeii or elsewhere, which confirmed in the most striking manner all that was written in the Gospels. But I found it more and more difficult, with free scope given to my imagination, to invent evidence which would suffice to convince me. Thus disbelief crept over me at a very slow rate, but was at last complete. The rate was so slow that I felt no distress.

Although I did not think much about the existence of a personal God until a considerably later period of my life, I will here give the vague conclusions to which I have been driven. The old argument from design in Nature, as given by Paley, which formerly seemed to me so conclusive, fails, now that the law of natural selection has been discovered. We can no longer argue that, for instance, the beautiful hinge of a bivalve shell must have been made by an intelligent being, like the hinge of a door by man. There seems to be no more design in the variability of organic beings, and in the action of natural selection, than in the course which the wind blows. But I have discussed this subject at the end of my book on the "Variations of Domesticated Animals and Plants," and the argument there given has never, as far as I can see, been answered.

But passing over the endless beautiful adaptations which we everywhere meet with, it may be asked how can the generally beneficent arrangement of the world be accounted for? Some writers indeed are so much impressed with the amount of suffering in the world, that they doubt, if we look to all sentient beings, whether there is more of misery or of happiness; whether the world as a whole is a good or bad one. According to my judgment happiness decidedly prevails, though this would be very difficult to prove. If the truth of this conclusion be granted, it harmonizes well with the effects which we might expect from natural selection. If all the individuals of any species were habitually to suffer to an extreme degree, they would neglect to propagate their kind; but we have no reason to believe that this has ever, or at least often occurred. Some other considerations, moreover, lead to the belief that all sentient beings have been formed so as to enjoy, as a general rule, happiness.

Everyone who believes, as I do, that all the corporeal and mental organs (excepting those which are neither advantageous nor disadvantageous to the possessor) of all beings have been developed through natural selection, or the survival of the fittest, together with use or habit, will admit that these organs have been formed so that their possessors may compete successfully with other beings, and thus increase in number. Now an animal may be led to pursue that course of action which is most beneficial to the species by suffering, such as pain, hunger, thirst, and fear; or by pleasure, as in eating and drinking, and in the propagation of the species, etc.; or by both means combined, as in the search for food. But pain or suffering of any kind, if long continued, causes depression and lessens the power of action, yet is well adapted to make a creature guard itself against any great or sudden evil. Pleasurable sensations, on the other hand, may be long continued without any depressing effect; on the contrary, they stimulate the whole system to increased action. Hence it has come to pass that most or all sentient beings have been developed in such a manner, through natural selection, that pleasurable sensations serve as their habitual guides. We see this in the pleasure from exertion, even occasionally from great exertion of the body or mind,—in the pleasure of our daily meals, and especially in the pleasure derived from sociability, and from loving our families. The sum of such pleasures as these, which are habitual or frequently recurrent, give, as I can hardly doubt, to most sentient beings an excess of happiness over misery, although many occasionally suffer much. Such suffering is quite compatible with the belief in Natural Selection, which is not perfect in its action, but tends only to render each species as successful as possible in the battle for life with other species, in wonderfully complex and changing circumstances.

That there is much suffering in the world no one disputes. Some have attempted to explain this with reference to man by imagining that it serves for his moral improvement. But the number of men in the world is as nothing compared with that of all other sentient beings, and they often suffer greatly without any moral improvement. This very old argument from the existence of suffering against the existence of an intelligent First Cause seems to me a strong one; whereas, as just remarked, the presence of much suffering agrees well with the view that all organic beings have been developed through variation and natural selection.

At the present day the most usual argument for the existence of an intelligent God is drawn from the deep inward conviction and feelings which are experienced by most persons.

Formerly I was led by feelings such as those just referred to (although I do not think that the religious sentiment was ever strongly developed in me), to the firm conviction of the existence of God, and of the immortality of the soul. In my Journal I wrote that whilst standing in the midst of the grandeur of a Brazilian forest, "it is not possible to give an adequate idea of the higher feelings of wonder, admiration, and devotion, which fill and elevate the mind." I well remember my conviction that there is more in man than the mere breath of his body. But now the grandest scenes would not cause any such convictions and feelings to rise in my mind. It may be truly said that I am like a man who has become colour-blind, and the universal belief by men of the existence of redness makes my present loss of perception of not the least value as evidence. This argument would be a valid one if all men of all races had the same inward conviction of the existence of one God; but we know that this is very far from being the case. Therefore I cannot see that such inward convictions and feelings are of any weight as evidence of what really exists. The state of mind which grand scenes formerly excited in me, and which was intimately connected with a belief in God, did not essentially differ from that which is often called the sense of sublimity; and however difficult it may be to explain the genesis of this sense, it can hardly be advanced as an argument for the existence of God, any more than the powerful though vague and similar feelings excited by music.

With respect to immortality, nothing shows me [so clearly] how strong and almost instinctive a belief it is, as the consideration of the view now held by most physicists, namely that the sun with all the planets will in time grow too cold for life, unless indeed some great body dashes into the sun, and thus gives it fresh life. Believing as I do that man in the distant future will be a far more perfect creature than he now is, it is an intolerable thought that he and all other sentient beings are doomed to complete annihilation after such long-continued slow progress. To those who fully admit the immortality of the human soul, the destruction of our world will not appear so dreadful.

Another source of conviction in the existence of God, connected with the reason, and not with the feelings, impresses me as having much more weight. This follows from the extreme difficulty or rather impossibility of conceiving this immense and wonderful universe, including man with his capacity of looking far backwards and far into futurity, as the result of blind chance or necessity. When thus reflecting I feel compelled to look to a First Cause having an intelligent mind in some degree analogous to that of man; and I deserve to be called a Theist. This conclusion was strong in my mind about the time, as far as I can remember, when I wrote the "Origin of Species"; and it is since that time that it has very gradually, with many fluctuations, become weaker. But then arises the doubt, can the mind of man, which has, as I fully believe, been developed from a mind as low as that possessed by the lowest animals, be trusted when it draws such grand conclusions?

I cannot pretend to throw the least light on such abstruse problems. The mystery of the beginning of all things is insoluble by us; and I for one must be content to remain an Agnostic.

When Bertrand Russell was sent to prison for opposing England's entrance into the first World War, the warden asked him what his religion was. Russell replied, "Agnostic." After asking Russell how to spell it, the warden sighed and said, "Well, there are many religions, but I suppose they all worship the same God." "This remark," Russell adds in his autobiography, "kept me cheerful for about a week."

CHAPTER 26

The Wandering Jew

(This essay first appeared in Free Inquiry, *Summer 1995.)*

> For the son of man shall come in the glory of his Father, with his
> angels; and then he shall reward every man according to his
> works. Verily I say unto you, There be some standing here, which
> shall not taste of death, till they see the Son of man coming in
> his kingdom.
>
> —*Matthew 16: 27, 28*

The statement of Jesus quoted above from Matthew, and repeated in similar words by Mark (8:38, 9:1) and Luke (9:26, 27), is for Bible
fundamentalists one of the most troublesome of all New Testament passages.

It is possible, of course, that Jesus never spoke those sentences, but all
scholars agree that the first-century Christians expected the Second Coming in their lifetimes. In Matthew 24, after describing dramatic signs of his
imminent return, such as the falling of stars and the darkening of the
moon and sun, Jesus added: "Verily I say unto you. This generation shall
not pass until all these things be fulfilled."

Until about 1933, Seventh-day Adventists had a clever way of rationalizing this prophecy. They argued that a spectacular meteor shower of 1833 was the falling of the stars, and that there was a mysterious darkening of sun and moon in the United States in 1870. Jesus meant that a future generation witnessing these celestial events would be the one to experience his Second Coming.

For almost a hundred years, Adventist preachers and writers of books assured the world that Jesus would return within the lifetimes of some who had seen the great meteor shower of 1833. After 1933 passed, the church gradually abandoned this interpretation of Jesus' words. Few of today's faithful are even aware that their church once trumpeted such a view. Although Adventists still believe Jesus will return very soon, they no longer set conditions for an approximate date.

How do they explain the statements of Jesus quoted in the epigraph? Following the lead of Saint Augustine and other early Christian commentators, they take the promise to refer to Christ's Transfiguration. Ellen White, the prophetess who with her husband founded Seventh-day Adventism, said it this way in her life of Jesus, *The Desire of Ages:* "The Savior's promise to the disciples was now fulfilled. Upon the mount the future kingdom of glory was represented in miniature. . . ."

Hundreds of Adventist sects since the time of Jesus, starting with the Montanists of the second century, have all interpreted Jesus' prophetic statements about his return to refer to *their* generation. Apocalyptic excitement surged as the year 1000 approached. Similar excitement is now gathering momentum as the year 2000 draws near. Expectation of the Second Coming is not confined to Adventist sects. Fundamentalists in mainstream Protestant denominations are increasingly stressing the imminence of Jesus's return. Baptist Billy Graham, for example, regularly warns of the approaching battle of Armageddon and the appearance of the Antichrist. He likes to emphasize the Bible's assertion that the Second Coming will occur after the gospel is preached to all nations. This could not take place, Graham insists, until the rise of radio and television.

Preacher Jerry Falwell is so convinced that he will soon be raptured—caught up in the air to meet the return of Jesus—that he once said he has no plans for a burial plot. Austin Miles, who once worked for Pat Robertson, reveals in his book *Don't Call Me Brother* (1989) that Pat once seriously considered plans to televise the Lord's appearance in the skies! Today's top native drumbeater for a soon Second Coming is Hal Lindsay. His many books on the topic, starting with *The Late Great Planet Earth,* have sold by the millions.

For the past two thousand years, individuals and sects have been setting dates for the Second Coming. When the Lord fails to show, there is often no recognition of total failure. Instead, errors are found in the calculations and new dates set. In New Harmony, Indiana, an Adventist sect called the Rappites was established by George Rapp. When he became ill he said that were he not absolutely certain the Lord intended him and his flock to witness the return of Jesus, he would think this was his last hour. So saying, he died.

The Catholic Church, following Augustine, long ago moved the Second Coming far into the future at some unspecified date. Liberal Protestants have tended to take the Second Coming as little more than a metaphor for the gradual establishment of peace and justice on earth. Julia Ward Howe, a Unitarian minister, had this interpretation in mind when she began her famous "Battle Hymn of the Republic" with "Mine eyes have seen the glory of the coming of the Lord. . . ." Protestant fundamentalists, on the other hand, believe that Jesus described actual historical events that would precede his literal return to earth to banish Satan and judge the quick and the dead. They also find it unthinkable that the Lord could have blundered about the time of his Second Coming.

The difficulty in interpreting Jesus' statement about some of his listeners not tasting of death until he returned is that he described the event in exactly the same phrases he used in Matthew 24. He clearly was not there referring to his transfiguration, or perhaps (as another "out" has it) to the fact that his kingdom would soon be established by the formation of the early church. Assuming that Jesus meant exactly what he said, and that he was not mistaken, how can his promise be unambiguously justified?

During the Middle Ages, several wonderful legends arose to preserve the accuracy of Jesus' prophecies. Some were based on John 21. When Jesus said to Peter, "Follow me," Peter noticed John walking behind him and asked, "Lord, what shall this man do?" The Lord's enigmatic answer was, "If I will that he tarry till I come, what is that to thee?"

We are told that this led to a rumor that John would not die. However, the writer of the fourth gospel adds: "Yet Jesus said not unto him, He shall not die; but if I will that he tarry till I come, what is that to thee?" Theologians in the Middle Ages speculated that perhaps John did not die. He was either wandering about the earth, or perhaps he ascended bodily into heaven. A more popular legend was that John had been buried in a state of suspended animation, his heart faintly throbbing, to remain in this unknown grave until Jesus returns.

These speculations about John rapidly faded as a new and more powerful legend slowly took shape. Perhaps Jesus was not referring to John when he said he could ask someone to tarry, but to someone else. This would also explain the remarks quoted in the epigraph. Someone not mentioned in the gospels, alive in Jesus' day, was somehow cursed to remain alive for centuries until judgment day, wandering over the earth and longing for death.

Who was this Wandering Jew? Some said it was Malchus, whose ear Peter sliced off. Others thought it might be the impenitent thief who was crucified beside Jesus. Maybe it was Pilate, or one of Pilate's servants. The version that became dominant identified the Wandering Jew as a shopkeeper—his name varied—who watched Jesus go by his doorstep, staggering under the weight of the cross he carried. Seeing how slowly and painfully the Lord walked, the man struck Jesus on the back, urging him to go faster. "I go," Jesus replied, "but you will tarry until I return."

As punishment for his rudeness, the shopkeeper's doom is to wander the earth, longing desperately to die but unable to do so. In some versions of the legend, he stays the same age. In others, he repeatedly reaches old age only to be restored over and over again to his youth. The legend seems to have first been recorded in England in the thirteenth century before it rapidly spread throughout Europe. It received an enormous boost in the early seventeenth century when a pamphlet appeared in Germany about a Jewish shoemaker named Ahasuerus who claimed to be the Wanderer. The pamphlet was endlessly reprinted in Germany and translated into other languages. The result was a mania comparable to today's manias for seeing

UFOs, Abominable Snowmen, and Elvis Presley. Scores of persons claiming to be the Wandering Jew turned up in cities all over England and Europe during the next two centuries. In the United States as late as 1868 a Wandering Jew popped up in Salt Lake City, home of the Mormon Adventist sect. It is impossible now to decide in individual cases whether these were rumors, hoaxes by impostors, or cases of self-deceived psychotics.

The Wandering Jew became a favorite topic for hundreds of poems, novels, and plays, especially in Germany, where such works continue to proliferate to this day. Even Goethe intended to write an epic about the Wanderer, but he only finished a few fragments. It is not hard to understand how anti-Semites in Germany and elsewhere would see the cobbler as representing all of Israel, its people under God's condemnation for having rejected his son as their Messiah.

Gustave Doré produced twelve remarkable woodcuts depicting episodes in the Wanderer's life. They were first published in Paris in 1856 to accompany a poem by Pierre Dupont. English editions followed with translations of the verse.

By far the best-known novel about the Wanderer is Eugene Sue's French work *Le Juif Errant (The Wandering Jew),* first serialized in Paris in 1844–45 and published in ten volumes. George Croly's three-volume *Salathiel* (1927, later retitled *Tarry Thou Till I Come*) was an enormously popular earlier novel. (In *Don Juan,* Canto 11, Stanza 57, Byron calls the author "Reverend Roley-Poley.") In Lew Wallace's *Prince of India* (1893), the Wanderer is a wealthy Oriental potentate.

George Macdonald's *Thomas Wingfold, Curate* (1876) introduces the Wandering Jew as an Anglican minister. Having witnessed the Crucifixion, and in constant agony over his sin, Wingfold is powerless to overcome a strange compulsion. Whenever he passes a roadside cross, or even a cross on top of a church, he has an irresistible impulse to climb on the cross, wrap his arms and legs around it, and cling there until he drops to the ground unconscious! He falls in love, but, realizing that his beloved will age and die while he remains young, he tries to kill himself by walking into an active volcano. His beloved follows, but is incinerated by the molten lava. There is a surprisingly happy ending. Jesus appears, forgives the Wanderer, and leads him off to paradise to reunite with the woman who died for him. The novel is not among the best of this Scottish writer's many admired fantasies.

My First Two Thousand Years, by George Sylvester Viereck and Paul Eldridge (1928), purports to be the erotic autobiography of the Wandering Jew. The same two authors, in 1930, wrote *Salome, the Wandering Jewess,* an equally erotic novel covering her two thousand years of lovemaking. The most recent novel about the Wanderer is by German ex-Communist Stefan Heym, a pseudonym for Hellmuth Flieg. In his *The Wandering Jew,* published in West Germany in 1981 and in a U.S. edition three years later, the Wanderer is a hunchback who tramps the roads with Lucifer as his companion. The fantasy ends with the Second Coming, Armageddon, and the Wanderer's forgiveness.

Sue's famous novel is worth a quick further comment. The Wanderer is Ahasuerus, a cobbler. His sister Herodias, the wife of King Herod, becomes the Wandering Jewess. The siblings are minor characters in a complex plot. Ahasuerus is tall, with a single black eyebrow stretching over both eyes like a Mark of Cain. Seven nails on the soles of his iron boots produce crosses when he walks across snow. Wherever he goes, an outbreak of cholera follows. Eventually the two siblings are pardoned and allowed "the happiness of eternal sleep." Sue was a French socialist. His Wanderer is a symbol of exploited labor, Herodias a symbol of exploited women. Indeed, the novel is an angry blast at Catholicism, capitalism, and greed.

The Wandering Jew appears in several recent science fiction novels, notably Walter Miller's *A Canticle for Leibowitz* (1959) and Wilson Tucker's *The Planet King* (1959), in which he becomes the last man alive on earth. At least two movies have dealt with the legend, the most recent a 1948 Italian film starring Vittorio Gassman.

Rafts of poems by British and U.S. authors have retold the legend. The American John Saxe, best known for his verse about the blind men and the elephant, wrote a seventeen-stanza poem about the Wanderer. British poet Caroline Elizabeth Sarah Norton's forgettable "Undying One" runs to more than a hundred pages. Oliver Herford, an American writer of light verse, in "Overheard in a Garden" turns the Wanderer into a traveling salesman peddling a book about himself. "The Wandering Jew" (1920) by Edwin Arlington Robinson is surely the best of such poems by an American writer.

In England, Shelley was the most famous poet to become fascinated by the legend. In his lengthy poem "The Wandering Jew," written or partly written when he was seventeen, the Wanderer is called Paulo. He attempts to conceal a fiery cross on his forehead under a cloth band. In the third canto, after sixteen centuries of wandering, Paulo recounts the origin of his suffering to Rosa, a woman he loves.

> *How can I paint that dreadful day,*
> *That time of terror and dismay,*
> *When, for our sins, a Saviour died,*
> *And the meek Lamb was crucified!*
> *As dread that day, when, borne along*
> *To slaughter by the insulting throng,*
> *Infuriate for Deicide,*
> *I mocked our Savior, and I cried,*
> *"Go, go," "Ah! I will go," said he,*
> *"Where scenes of endless bliss invite;*
> *To the blest regions of the light*
> *I go, but thou shalt here remain—*
> *Thou diest not till I come again."—*

The Wandering Jew is also featured in Shelley's short poem "The Wandering Jew's Soliloquy," and in two much longer works, "Hellas" and "Queen Mab." In "Queen Mab," as a ghost whose body casts no shadow, Ahasuerus bitterly denounces God as an evil tyrant. In a lengthy note about this, Shelley quotes from a fragment of a German work "whose title I have vainly endeavored to discover. I picked it up, dirty and torn, some years ago. . . ."

In this fragment the Wanderer describes his endless efforts to kill himself. He tries vainly to drown. He leaps into an erupting Mount Etna, where he suffers intense heat for ten months before the volcano belches him out. Forest fires fail to consume him. He tries to get killed in wars, but arrows, spears, clubs, swords, bullets, mines, and trampling elephants have no effect on him. "The executioner's hand could not strangle me . . . nor would the hungry lion in the circus devour me." Snakes and dragons are powerless to harm him. He calls Nero a "bloodhound" to his face, but the tyrant's tortures cannot kill him.

> Ha! not to be able to die—not to be able to die—not to be permitted to rest after the toils of life—to be doomed to be imprisoned forever in the clay-formed dungeon—to be forever clogged with this worthless body, its load of diseases and infirmities—to be condemned to hold for millenniums that yawning monster Sameness, and Time, that hungry hyena, ever bearing children and ever devouring again her offspring! Ha! not to be permitted to die! Awful avenger in heaven, hast thou in thine army of wrath a punishment more dreadful? then let it thunder upon me; command a hurricane to sweep me down to the foot of Carmel that I there may lie extended; may pant, and writhe, and die!

Scholarly histories of the legend have been published in Germany and elsewhere. In English, Moncure Daniel Conway's *The Wandering Jew* (1881) has become a basic reference. See also his article on the Wanderer in the *Encyclopaedia Britannica's* ninth edition. Another valuable account is given by Sabine-Baring Gould in his *Curious Myths of the Middle Ages* (second edition, 1867).

The definitive modern history is George K. Anderson's *The Legend of the Wandering Jew,* published by Brown University Press in 1965. A professor of English at Brown, Anderson made good use of the university's massive collection of literature about the Wanderer. His book's 489 pages contain excellent summaries of European poems, plays, and novels not touched upon here, as well as detailed accounts of the many claimants. The book may tell you more than you care to know about this sad attempt of Christians to avoid admitting that the Galilean carpenter turned preacher did indeed believe he would soon return to earth in glory, but was mistaken.

Addendum

In Italy the legend of the Wandering Jew took a charming and completely different form. Befana was sweeping her house when the three Wise Men rode by and invited her to go with them to Bethlehem. Befana said she was much too busy. Later, regretting her decision, she began wandering about the world under a terrible curse that does not allow her to die. Each year on the eve of Twelfth Night (January 5), a day that commemorates the visit of the Magi, Befana slides down the chimney on her broom to fill shoes and stockings with candy and small toys. She always peers into the faces of the sleeping children, hoping to see the infant Jesus.

Befana's story is told in the following doggerel. I found it in *The Peerless Speaker* (1900), where it is credited to Louise V. Boyd.

> *"COME forth, come forth, Beffana!"*
>   *She hears her neighbors say,*
> *"Come, up the road to Bethlehem*
>   *The Wise Men pass to-day!"*
>
> *So busy was Beffana*
>   *She scarcely turned her head;*
> *Here was the waiting linen,*
>   *The waiting scarlet thread.*
>
> *Again they cried, "Beffana,*
>   *It is a glorious sight,*
> *Three Kings together journey*
>   *In crowns and garments bright!"*
>
> *Her people's skillful daughters*
>   *As yet she had excelled.*
> *Beffana saw the spindle,*
>   *Her hand the distaff held;*

Her husband's words must praise her.
  Her children's voices bless;
She eateth in her household
  No bread of idleness.

So she made haste to answer,
  "My house is all my care;
No time have I for strangers
  Toward Bethlehem that fare!

"Ere yet the daytime cometh
  I give my household meat:
Mine is the best-clad husband
  That hath an elder's seat.

"And merchants know my girdles
  And my woven tapestry,
The glory of my purple
  And silk most fair to see!"

But now her kinsmen shouted,
  "You know not what you miss!
There may be many pageants,
  Yet none be like to this!

"Men say the three Kings journey
  A wondrous thing to see,
A babe born of a Virgin
  Foretold by prophecy.

"Oh! come: behold, Beffana!
  For speech may never say
The splendor on their faces,
  The Kings that ride this way!"

Beffana still kept busy,
   But lightly answered then:
"I will look out upon them
   As they come back again!"

But all her friends and kinsmen,
   In wondering delight,
Gazed till the Kings so gently
   Had journeyed out of sight.

That eve Beffana's husband
   Had sorrow in his gaze,
When of her work she told him,
   Anticipating praise.

He did not quite upbraid her,
   But out of ancient lore
He questioned, "Who hath profit
   In laboring evermore?"

And spake of times for mourning
   And times to laugh and sing;
Of times to keep or scatter,
   Of times for everything.

And, sad, Beffana answered:
   "My lord is right, but then
I surely will behold them
   As they come back again."

Alas! alas! Beffana
   Looked out from day to day,
They came no more; God warned them
   To go another way.

And she grew very weary
    Who had so much to do,
And never came the vision
    That might her strength renew.

Beffana dieth never,
    This earth is still her home;
Beffana looketh ever
    For those who never come.

Many old anthologies contain the following sad poem about the Wandering Jew, translated from the German by Charles Timothy Brooks, a nineteenth-century Unitarian minister, poet, and translator.

THE *Wandering Jew once said to me,*
    I passed through a city in the cool of the year,
A man in the garden plucked fruit from a tree;
    I asked, "How long has this city been here?"
And he answered me, and he plucked away,
"It has always stood where it stands to-day,
And here it will stand forever and aye."
    Five hundred years rolled by, and then
    I travelled the self-same road again.

No trace of a city there I found;
    A shepherd sat blowing his pipe alone,
His flock went quietly nibbling round,
    I asked, "How long has the city been gone?"
And he answered me, and he piped away,
"The new ones bloom and the old decay,
This is my pasture ground for aye."
    Five hundred years rolled by, and then
    I travelled the self-same road again.

And I came to a sea, and the waves did roar,
And a fisherman threw his net out clear,
And when heavy laden he dragged it ashore,
I asked, "How long has the sea been here?"
And he laughed, and he said, and he laughed away:
"As long as yon billows have tossed their spray,
They've fished and they've fished in the self-same way."
Five hundred years rolled by, and then
I travelled the self-same road again.

And I came to a forest, vast and free,
And a woodman stood in the thicket near;
His axe he laid at the foot of a tree:
I asked, "How long have the woods been here?"
And he answered, "The woods are a covert for aye;
My ancestors dwelt here alway,
And the trees have been here since creation's day."
Five hundred years rolled by, and then
I travelled the self-same road again.

And I found there a city, and far and near
Resounded the hum of toil and glee,
And I asked, "How long has the city been here,
And where is the pipe, and the wood, and the sea?"
And they answered me, and they went their way,
"Things always have stood as they stand to-day,
And so they will stand for ever and aye."
I'll wait five hundred years, and then
I'll travel the self-same road again.

The Wandering Jew, by Gustav Doré

CHAPTER 27

The Second Coming

As the year 2000 approached, Protestant fundamentalists (I include members of Pentecostal churches and such fringe sects as Seventh-day Adventism and Jehovah's Witnesses) became more and more persuaded that the Lord's Second Coming was close at hand. Scores of strident books were published, and are still being published, showing how a correct interpretation of the books of Daniel and Revelation proves that the rapture of believers, the Battle of Armageddon, and the end of the world as we know it will be occurring very, very soon. The books range from the many by Hal Lindsey, which have sold by the millions, to obscure volumes which identify the Antichrist and reveal the meaning of 666, his number.

You would think that believers in the imminence of Christ's return would be bothered by the fact that, ever since the gospels were written, huge numbers of Christians have interpreted Biblical signs of the end as applying to *their* generation. The sad history of these failed prophecies makes no impression on the mind-sets of today's fundamentalists. Even Billy Graham, who should know better, has for decades preached and written about the impending return of Jesus. He grants that no one knows the exact year, but all signs indicate, he believes, that the great event is almost upon us.

It is often said that current excitement over the Second Coming, centering on the year 2000, had its parallel in a panic over the end of the world that swept through Christian Europe as the year 1000 approached. But did such panic actually occur? As Stephen Jay Gould makes clear in his wise little book *Questioning the Millennium* (1997), the answer is far from clear. There is now, he tells us, an enormous literature on the topic that spans the full range of opinion from the claim that Europe did indeed experience "panic terror" to the claim that nothing of the sort took place.

Gould cites Richard Erdoes's *AD 1000: Living on the Brink of the Apocalypse* (1988) as a recent defense of the panic terror school. A German now living in Santa Fe, New Mexico, Erdoes is the author of two previous books, *The Sundance Principle* and *American Indian Myths*. "On the last day of the year 999," Erdoes begins his history, ". . . the old Basilica of St. Peter's at Rome was thronged with a mass of weeping and trembling worshippers awaiting the end of the world."

At the other end of the spectrum, Gould cites *Century's End* (1990), by Hillel Schwartz. Schwartz denies that any undue excitement over the Second Coming took place as 1000 loomed. An intermediate view, that there was *some* excitement but not much, is ably championed by French historian Henry Focillon in *The Year 1000* (English translation, 1969), and by Richard Landes, a Boston University historian. In *Visions of the End: Apocalyptic Traditions in the Middle Ages* (Harvard University Press, 1995), Landes points out that because the world failed to end there was no desire to keep records. He thinks it insulting to suppose that the populace did not know the year was 1000. For a good account of this debate, see Patricia Bernstein's article, "Terror in A.D. 1000," in *Smithsonian* (July 1999, pp. 115–25), and the references she cites.

Gould admits that he favored Schwartz's position until he attended an international conference devoted to "The Apocalyptic Year 1000," held at Boston University in 1996. The conference organizer, medieval historian Richard Landes, convinced Gould that there was considerable "millennial stirring" in the year 1000, especially among European peasants. One major drum beater for millennial terror was a monk named Raoul Glaber. Like almost all such failed prophets, Glaber found an error in his calculations when Christ did not appear. The thousand years, he proclaimed, should not be counted after Christ's birth, but after his death. This postponed the world's end, he said, until 1033.

Hundreds of predictions have been made around the world as the year 2000 approached, about the date of the Lord's return. Here are some recent examples that are especially comic.

In 1988 Edgar C. Whisenant, then fifty-six, a retired NASA rocket engineer living in Little Rock, Arkansas, published a paperback booklet titled *88 Reasons Why the Rapture Will Be in 88*. The publisher, a firm in Santa Rosa, California, claimed they sold or gave away over six million copies. The book predicted that the rapture would occur on September 11, 12, or 13, 1988. When the event failed to take place, Whisenant found a slight error in his calculations, and moved the date ahead to September 1, 1989. When *that* date also proved wrong, Whisenant decided henceforth to keep his mouth shut. He told a reporter he was under medication to control paranoid schizophrenia, but that his mental condition had no bearing on his calculations.

Robert W. Faid's *Gorbachev! Has the Real Antichrist Come?* was published in 1988 by Victory House, a fundamentalist firm in Tulsa. Faid is identified on the cover as a nuclear engineer and author of *A Scientific Approach to Christianity.* He lives in Taylors, South Carolina. Using elaborate systems of numerology, Faid finds that in one system Gorbachev's full name yields 666, and in another system it produces 888, a number Faid identifies with Jesus. Gorbachev is thus shown to be both the Beast of Revelation and the counterfeit Christ. The Second Coming, Faid warns, will take place in 2000 or shortly thereafter. A portion of his crazy book was actually reprinted in *Harper's Magazine* (January 1989). I have no idea whether Faid today still thinks poor Gorby is the incarnation of Satan.

Correspondent John Earwood called my attention to a much funnier book. Titled *666: The Final Warning,* the author is Gary D. Blevins, a former Prudential Life Insurance agent, now a financial consultant in Tennessee. This lavishly illustrated paperback was privately published in 1990 by Blevins's Visions of the End Ministries, and can be obtained by writing to P.O. Box 944, Kingston, TN 37662. The book has 494 pages and an introduction by Texe Marrs, another fundamentalist, and author of several bestselling books.

Blevins's book is based throughout on what he calls the Bible's Secret Code, a code concocted by other fundamentalists whose books he recommends. The code is simple. Each letter is assigned a number that is the product of 6 and the letter's position in the alphabet. Thus $A = 1 \times 6 = 6$, $B = 2 \times 6 = 12$, $C = 3 \times 6 = 18$, and so on to $Z = 6 \times 26 = 156$.

Blevins must have labored long and hard at his calculations, applying the code to hundreds of names and phrases to produce relevant sums, and especially the sum of 666, Revelation's notorious "number of the Beast."

Blevins writes that he was surprised to find that *Kissinger* adds to 666, but he realized at once that Henry Kissinger couldn't be the Antichrist because he failed to fit "Scripture guidelines." He was also amazed that so many common words and phrases, such as *New York, illusion, witchcraft, necromancy, Mark of Beast,* and *Santa Claus* add to 666.

If not *Kissinger,* then who *does* Blevins think, or perhaps I had best say *thought* in 1990, is the primary suspect for being the Antichrist? You won't believe it, but the candidate is none other than Ronald Wilson Reagan!

Each of Reagan's three names has six letters, and the entire name has six syllables. This is suspicious enough, but Blevins is compelled to do more. Unfortunately *Ronald Reagan* is six short of 666, but Blevins remedies this by adding *A* in front of the name: *A Ronald Reagan.* That's not all. A tireless Blevins manages to find scores of other phrases about Reagan that add to 666. Here are some of them:

Office of Reagan, Rank of Reagan, A Mark of Reagan, Space of Reagan, Ray of Reagan, Vim of Reagan, Tact of Reagan, Talk of Reagan, Brain of Reagan, Mold of Reagan, Peer of Reagan, Karma of Reagan, Ranch of Reagan, Hope of Reagan, Faith of Reagan, Old Age of Reagan, Creme of Reagan, Reagan in Japan, and dozens of other phrases.

One might object that even in 1990, when Blevins's book was published, Reagan was no longer in power. This doesn't faze Blevins one bit. Does not Revelation 17:8 speak of "the beast that was, and is not, and yet is?" To Blevins this tells us that Reagan will regain power, but now on a global scale. He will rule the world by means of a supercomputer (Blevins's code gives to *computer* a sum of 666), and by keeping track of everybody with bar codes implanted in hands and foreheads. He will be assisted by the Masons (Blevins believes Freemasonry is a satanic cult) and by the present Pope. Blevins reminds us that Reagan is an honorary Mason, that he believes in astrology and lucky charms, and that 33 is his lucky number.

Blevins allows that he is not absolutely certain that Reagan is destined to become the Beast, he says he likes Reagan personally, and hopes Reagan will not turn out to be the Antichrist. However, "the alarm must be sounded." In Blevins's opinion the evidence is "overwhelming" that Reagan is the prime suspect.

Blevins provides a tentative outline of what the next few years have in store. In 1991–94 New York City will be destroyed and UFOs will land. In 1996 Reagan's mind, invaded by Satan, will be transformed into the Antichrist, who will rule the world for a thousand years. In 1998 Reagan will be cast into the Lake of Fire, the faithful will be raptured, Jesus will come back, and Satan will be bound for a thousand years. In 3000 Satan will go into the Lake of Fire along with the resurrected unsaved, and Jesus will rule over a peaceful new Earth.

"Most real theologians in our day," Blevins writes, "flatly state that we will not see the year 2000 before the Lord returns! I strongly agree with that statement."

Now that 1998 has passed with no sign of the Lord, and Reagan surely is no longer capable of ruling the world, one would suppose that an embarrassed Blevins would apologize for his blunders and withdraw his book from the market. But no. In 1999 I sent him $16.50 for a copy. It arrived promptly with nary a hint of a disclaimer. Blevins's Vision of the End Ministries must need the money.

In Seoul, South Korea, in 1992, Lee Jang Rim, head of one of some two hundred Protestant churches in that country, created nationwide hysteria by announcing that the rapture would take place on October 28, 1992. The prophecy was based on a vision that came to a sixteen-year-old boy. Twenty thousand Korean fundamentalists in South Korea, Los Angeles, and New York City took the prediction seriously. Hundreds quit jobs, left families, and had abortions to prepare for their trip to heaven. Rim's church paid for costly ads in the *Los Angeles Times* and the *New York Times*. They urged readers to prepare for their journey through the skies, and to refuse to allow 666 to be imprinted in bar code on their forehead or right hand.

Riot police, plainclothes officers, and reporters crowded outside Korean churches, flanked by fire engines, ambulances, and searchlights. Believers took the failure of the prophecy calmly, and there were no reported riots. Only sadness. In December 1992 Rim was arrested and sentenced to two years in prison for having bilked $4.4 million from his flock. He had invested the money in bonds that didn't mature until the following year!

In 1992 Harold Camping published, through a vanity press, his book *1994?* It predicted that the Second Coming would occur in September of that year. This was followed in 1993 by a sequel titled *Are You Ready?* Together, the two books total 955 pages. Trained as a civil engineer, Camping made enough money running a construction company to found, in 1959, Family Stations, Inc. It soon came to control thirty-nine radio stations. A nonordained Bible scholar, Camping conducted a nightly radio talk show from his headquarters in Oakland, California. After September passed with no sign of the Lord, Camping changed his date to October 2. When that passed uneventfully, he ran out of excuses and decided against any more date setting.

Among Protestant sects the Seventh-day Adventists continue to be the most vocal predictors of an impending Second Coming, though they no longer set a date for that event. The church had its origin in the teachings of a simple-minded farmer named William Miller. His study of the Bible convinced him that 1843 would be the year Jesus would return. When this didn't happen he moved the date to October 22, 1844. After that prediction also failed, Miller had the good sense to stop predicting, but the undaunted Millerites decided that October 22, 1845, was the correct date. This was later moved ahead to 1851. After that year Adventist leaders wisely realized that such date setting was giving the sect a bad reputation.

In Matthew 24 Jesus describes the darkening of the sun and moon, and a falling of stars from the sky, as signs of his approaching return. "Verily I say unto you, this generation shall not pass till all these things be fulfilled."

Liberal Bible scholars have long agreed that "this generation" refers to the generation of those listening to Jesus' words. Because he did not return in that generation, fundamentalists of all stripes have been forced to reinterpret Christ's remarks in less plausible ways. William Miller preached that the darkening of the sun and moon actually took place in 1780, and that the falling star prediction was fulfilled in 1833 by a dramatic shower of meteors. The generation witnessing these events, Miller maintained, would be the generation that would also see the Lord return in glory.

Until about 1933 Seventh-day Adventist literature defended these Millerite views. Adventist books included dramatic pictures of the dark day and the falling "stars." The church taught that Jesus would surely return within the lifetime of at least some who had witnessed the 1833 meteor shower. When it became embarrassingly obvious that this could not be, the church quietly dropped from its literature all references to the dark day and the falling stars.

I was therefore surprised when I read *The Coming Great Calamity,* by Adventist Marvin Moore, published by his church in 1997. Moore edits the Adventist periodical *Signs of the Times* and has written three previous books: *The Crisis of the End Times, The Antichrist and the New World,* and *Conquering the Dragon Within.*

Ellen White, the Adventist-inspired visionary and one of the faith's founders, defends Miller's views about the dark day and falling stars in her masterpiece *The Great Controversy Between Christ and Satan.* This is very painful now to conservative Adventists who are unable to admit that Mrs. White could be wrong about anything. How does Moore manage to defend Mrs. White? He argues that she was correct in seeing the dark day and the 1833 shower as fulfillments of Matthew 24, but they were only *partial* fulfillments. They tell us "that the time of the end had begun, not that it was about to end."

The complete fulfillments of Matthew 24, Moore reasons, will be soon, with Earth's destruction caused by "comets, asteroids, and/or meteors." He admits he could be wrong, nevertheless he is convinced that the new millennium will undoubtedly be the century in which stars will seem to fall, the sun and moon will be obscured, and the Lord will return. Before he returns, Earth will experience a terrible destruction not seen since the great flood in the days of Noah.

Jehovah's Witnesses have an even worse record of failed predictions than the Adventists. They teach that Jesus returned in 1914, but it was an invisible, spiritual return. However, they also once taught that 1914 would see the beginning of Armageddon, followed by the destruction of all nations and the establishment of God's Kingdom on Earth. When this didn't happen, the date was moved to 1915. After that year passed, the date was pushed ahead again to 1918. Unfazed by the 1918 failure, 1975 was the next selection.

As far as I know, since then the group has stopped proposing dates, although it still preaches that the end times are near and millions now living will never die. It's useless to bring all this up when a Witness knocks on your door because most Witnesses today are ignorant of their faith's bizarre history, or about the errors and sins of Charles Taze Russell, who founded their sect. A good reference on the history of Jehovah's Witnesses is an article in the *Dictionary of Cults, Sects, Religions, and the Occult* (1993), by George A. Mather and Larry A. Nichols, and the many references they cite.

Addendum

Finding 666 in the names of famous people is a number-twiddling pastime that has obsessed numerologists ever since the book of Revelation was written. With patience and ingenuity it is not difficult to extract 666 from almost any person's name. For example, using Blevins's Bible code I discovered that *Sun Moon* and *Pat J. Buchanan* each adds to 666. The same code yields 666 if you apply it to *Hal Lindsey B,* the B standing, of course, for Beast.

My favorite candidate for the Antichrist is Jesse Ventura, former wrestling beast and now governor of Minnesota. Apply Blevins's code to *J. Ventura.* Bingo! 666.

Satan and *Beast* each have five letters. So let's start Blevins's code with A = 5, B = 6, and so on. Applied to *Blevins,* the code gives 666. Could Charlton Heston, chief spokesman for the gun lobby, be preparing the forces of evil for the Battle of Armageddon? Heston has six letters. If we number the alphabet A = 6, B = 7, and so on, then apply Blevins's technique of multiplying each value by six, *Heston* adds to 666.

With more effort I found a way to apply 666 to *Jerry Falwell.* Number the alphabet backward, starting with Z = 0, Y = 1, X = 2, and so on. I call this the Devil's Code. Take the values of the letters in *Falwell,* multiply each by 6, add, and you get 666. The Devil's Code also turns Billy Graham into the Antichrist if you write his name *W. Graham.*

Could President Clinton be the Antichrist? Add the normal position values of *W.J.C.,* the initials of William Jefferson Clinton, and you get 36. The sum of all numbers 1 through 36 is 666. I couldn't do much with *Castro,* but the same procedure produces 36 if you add the position values of *Fidel.*

Part X

The Last Word

CHAPTER 28

Science and the Unknowable

Existence, the preposterous miracle of existence! To whom has
the world of opening day never come as an unbelievable sight?
And to whom have the stars overhead and the hand and voice
nearby never appeared as unutterably wonderful, totally beyond
understanding? I know of no great thinker of any land or era
who does not regard existence as the mystery of all mysteries.
—*John Archibald Wheeler*

One of the fundamental conflicts in philosophy,
perhaps the most fundamental, is between those who believe that the universe open to our perception and exploration is all there is, and those who
regard the universe we know as an extremely small part of an unthinkably
vaster reality. These two views were taken by those two giants of ancient
Greek philosophy Aristotle and Plato. Plato, in his famous cave allegory,
likened the world we experience to the shadows on the wall of a cave. To
turn this into a mathematical metaphor, our universe is like a projection
onto three-dimensional space of a much larger realm in a higher space-time.

For Aristotle the universe we see, although parts of it are beyond human
comprehension, is everything. It is a steady-state cosmos, self-caused, hav-

ing no beginning or end. There is no Platonic realm of transcendent realities and deities. Plato succumbed to what Paul Kurtz likes to call the "transcendental temptation." Aristotle managed to avoid it.

In recent years cosmologists have blurred the distinction between the universe we know and transcendent regions by positing a "multiverse" in which an infinity of universes are continually exploding into existence, each with a unique set of laws and constants. This is one way to defend the anthropic principle against the argument that the universe's fine tuning is evidence of a Designer. It is known that if any of some dozen constants is altered by a minuscule fraction it would not be possible for suns and planets to form, let alone life to evolve. The counterargument: If there is an infinity of universes, each with an unplanned, random set of constants, then obviously we must exist only in a universe with constants that permit life to evolve.

The multiverse concept, however, is far from a step toward Platonic transcendence. The other universes do not differ from ours in any truly fundamental way. They all spring into being in response to random fluctuations in the same laws of quantum mechanics, varying only in the accidental way their big bang creates laws. There is still no need to leap from a godless nature to transcendental regions that somehow lie beyond the multiverse.

A few cosmologists and far-out philosophers have gone much further. They conjecture that all possible universes exist—that is, every universe based on a noncontradictory set of laws. In the many-worlds interpretation of quantum mechanics, the universe is constantly splitting into parallel worlds, but these countless worlds all obey the same laws. The multi-multiverse of all-possible-worlds is a much larger ensemble, obviously infinite because the number of logically consistent possibilities is infinite. Most physicists do not buy this view because it is the utmost imaginable violation of Occam's razor. Leibniz's notion of a Creator who surveyed all logically possible worlds, then selected what He considered the most desirable, is surely a simpler conjecture by many orders of magnitude.

A question now arises. As science steadily advances in its knowledge of nature, never reaching absolute certainty but always getting closer and closer to understanding nature, will it eventually discover everything?

We have to be careful to define what is meant by "everything." There is a trivial sense in which humanity cannot possibly know all there is to know. We will never know how many hairs were on Plato's head when he died, or whether Jesus sneezed while delivering the Sermon on the Mount. We will never know all the decimal digits of pi, or all possible theorems of geometry. We will never know all possible theorems just about triangles. We will never know all possible melodies, or poems, or novels, or paintings, or jokes, or magic tricks, because the possible combinations are limitless. Moreover, as Kurt Gödel taught us, every mathematical system complex enough to include arithmetic contains theorems that cannot be proved true or false within the system. Whether Gödelian undecidability may apply to mathematical physics is not yet known.

When physicists talk about TOEs (theories of everything) they mean something far less trivial. They mean that all the fundamental laws of physics eventually will become known, perhaps unified by a single equation or a small set of equations. If this happens, and physicists find what John Wheeler calls the Holy Grail, it will of course leave unknown billions and billions of questions about the complexities that emerge from the fundamental laws.

At the moment, cosmologists do not know the nature of "dark matter" that holds together galaxies, or how fast the universe is expanding, and there are hundreds of other unanswered questions. Biologists do not know how life arose on Earth or whether there is life on planets in other solar systems. Evolution is a fact, but deep mysteries remain about how it operates. No one has any idea how complex organic molecules are able to fold so rapidly into the shapes that allow them to perform their functions in living organisms. No one knows how consciousness emerges from the brain's complicated molecular structure. We do not even know how the brain remembers.

Such a list of unknowns could fill a book, but all of them are potentially knowable if humanity survives long enough. Too often in the past scientists have decided that something is permanently unknowable only to be contradicted a few generations later. On the other hand, many scientists have predicted that physics was near the end of its road only to have enormous new revolutions of knowledge take place a few decades later.

In recent years, just when it was thought that all the basic particles had been found or conjectured, along came superstrings, the most likely candidate at the moment for a TOE. If superstring theory is correct, it means that all fundamental particles are made of incredibly tiny loops of enormous tensile strength. The way they vibrate generates the entire zoo of particles.

What are superstrings made of? As far as anyone knows they are not made of anything. They are pure mathematical constructs. If superstrings are the end of the line, then everything that exists in our universe, including you and me, is a mathematical construction. As a friend once said, the universe seems to be made of nothing, yet somehow it manages to exist.

On the other hand, superstrings may, at some future time, turn out to be composed of still smaller entities. Many famous scientists, notably Arthur Stanley Eddington, David Bohm, Eugene Wigner, Freeman Dyson, and Stanislaw Ulam, believed that the universe has bottomless levels. As soon as one level is penetrated, a trapdoor opens to a hitherto unsuspected subbasement. These subbasements are infinite. As the old joke goes, it's turtles all the way down. Here is how Isaac Asimov expressed this opinion in his autobiography, *I, Asimov:* "I believe that scientific knowledge has fractal properties; that no matter how much we learn, whatever is left, however small it may seem, is just as infinitely complex as the whole was to start with. That, I think, is the secret of the Universe."

A similar infinity may go the other way. Our universe may be part of a multiverse, in turn part of a multi-multiverse, and so on without end. As one of H. G. Wells's fantasies has it, our cosmos may be a molecule in a ring on a gigantic hand.

Even if the universe is finite in both directions, and there are no other worlds, are there fundamental questions that can never be answered? The slightest reflection demands a yes.

Suppose that at some future date a TOE will provide all the basic laws and constants. Explanation consists of finding a general law that explains a fact or a less general law. Why does Earth go around the sun? Because it obeys the laws of gravity. Why are there laws of gravity? Because, Einstein revealed, large masses distort space-time, causing objects to move along geodesic paths. Why do objects take geodesic paths? Because they are the shortest paths through space-time. Why do objects take the shortest paths? Now we hit a stone wall. Time, space, and change are given aspects of reality. You can't define any of these concepts without introducing the concept into the definition, as physicists like to say, they are "incompressible" into more basic concepts. They are not mere aspects of human consciousness, as Kant imagined. They are "out there," independent of you and me. They may be unknowable in the sense that there is no way to explain them by embedding them in more general laws.

Imagine that physicists finally discover all the basic waves and their particles, and all the basic laws, and unite everything in one equation. We can then ask, "Why that equation?" It is fashionable now to conjecture that the big bang was caused by a random quantum fluctuation in a vacuum devoid of space and time. But of course such a vacuum is a far cry from nothing. There had to be quantum laws to fluctuate. And why are there quantum laws?

Even if quantum mechanics becomes "explained" as part of a deeper theory—call it X—as Einstein believed it eventually would be, then we can ask, "Why X?" There is no escape from the superultimate questions: Why is there something rather than nothing, and why is the something structured the way it is?" As Stephen Hawking recently put it, "Why does the universe go to all the bother of existing?" The question obviously can never be answered, yet it is not emotionally meaningless. Meditating on it can induce what William James called an "ontological wonder-sickness." Jean-Paul Sartre called it "nausea." Fortunately such reactions are short-lived or one could go mad by inhaling what James called "the blighting breath of the ultimate why."

Consider the extremely short time humanity has been evolving on our little planet. It seems unlikely that evolution has stopped with us. Can anyone believe that a million years from now, if humanity lasts, our brains will not have evolved far beyond their present capacities? Our nearest relatives, the chimpanzees, are incapable of understanding why three times three is nine, or anything else taught in grade school. It is difficult to imagine that a million years from now our brains will not be grasping truths about the universe that are as far beyond what we now can know as our understanding is beyond the mind of a monkey. To suppose that our brains, at this stage of an endless process of evolution, are capable of knowing everything that can be known strikes me as the ultimate in hubris.

If one is a theist, obviously there is a vast unknowable reality transcending our universe, a "wholly other" realm impossible to contemplate without an emotion of what Rudolph Otto called the *mysterium tremendum.* But even if one is an atheist or agnostic, the Unknowable will not go away. No philosopher has written more persuasively about this than agnostic Herbert Spencer in the opening chapters of his *First Principles* (1894).

In the beginning hundred pages of this book, in a part titled "The Unknowable," Spencer argues that a recognition of the Unknowable is the only way to reconcile science with religion. The emotion behind all religions, aside from their obvious superstitions and gross beliefs, is one of awe toward the impenetrable mysteries of the universe. Here is how Spencer reasoned:

One other consideration should not be overlooked—a consideration which students of Science more especially need to have pointed out. Occupied as such are with established truths, and accustomed to regard things not already known as things to be hereafter discovered, they are liable to forget that information, however extensive it may become, can never satisfy inquiry. Positive knowledge does not, and never can, fill the whole region of possible thought. At the uttermost reach of discovery there arises, and must ever arise, the question—What lies beyond? As it is impossible to think of a limit to space so as to exclude the idea of space lying outside that limit; so we cannot conceive of any explanation profound enough to exclude the question—What is the explanation of that explanation? Regarding Science as a gradually increasing sphere, we may say that every addition to its surface does but bring it into wider contact with surrounding nescience. There must ever remain therefore two antithetical modes of mental action. Throughout all future time, as now, the human mind may occupy itself, not only with ascertained phenomena and their relations, but also with that unascertained something which phenomena and their relations imply. Hence if knowledge cannot monopolize consciousness—if it must always continue possible for the mind to dwell upon that which transcends knowledge, then there can never cease to be a place for something of the nature of Religion; since Religion under all its forms is distinguished from everything else in this, that its subject matter passes the sphere of the intellect.

By "Religion" Spencer did not mean religion in the usual sense of worshiping God or gods, but only a sense of awe and wonder toward ultimate mysteries. For him Science and Religion were two essential aspects of thought, Science expressing the knowable, Religion the unknowable. The two merge without contradiction. "If Religion and Science are to be reconciled," he writes, "the basis of reconciliation must be this deepest, widest, most certain of all facts—that the Power which the Universe manifests to us is inscrutable."

No matter how many levels of generalization are made in explaining facts and laws, the levels must necessarily reach a limit beyond which science is powerless to penetrate.

In all directions his investigations eventually bring him face to face with an insoluble enigma; and he ever more clearly perceives it to be an insoluble enigma. He learns at once the greatness and the littleness of the human intellect—its power in dealing with all that comes within the range of experience, its impotence in dealing with all that transcends experience. He, more than any other, truly *knows* that in its ultimate nature nothing can be known.

The rest of Spencer's *First Principles,* titled "The Knowable," is an effort to summarize the science of his day, especially what was then known about evolution.

But an account of the Transformation of Things, given in the pages which follow, is simply an orderly presentation of facts; and the interpretation of the facts is nothing more than a statement of the ultimate uniformities they present—the laws to which they conform. Is the reader an atheist? The exposition of these facts and these laws will neither yield support to his belief nor destroy it. Is he a pantheist? The phenomena and the inferences as now to be set forth will not force on him any incongruous implication. Does he think that God is immanent throughout all things, from concentrating nebulae to the thoughts of poets? Then the theory to be put before him contains no disproof of that view. Does he believe in a Deity who has given unchanging laws to the Universe? Then he will find nothing at variance with his belief in an exposition of those laws and an account of the results.

Boundaries and Barriers: On the Limits of Scientific Knowledge (Addison-Wesley, 1996), edited by John Casti and Anders Karlqvist, is one of a spate of recent books on the topic. For almost all its authors, the term "limits" is confined to unsolved but potentially solvable questions. Most of the authors agree with what the editors say in their introduction: "Unlike mathematics, there is no knock-down airtight argument to believe that there are questions about the rest of the world that we cannot answer in principle."

Only British astronomer John Barrow has the humility to disagree. He concludes his contribution as follows:

> In this brief survey we have explored some of the ways in which the quest for a Theory of Everything in the third millennium might find itself confronting impassable barriers. We have seen there are limitations imposed by human intellectual capabilities, as well as by the scope of technology. There is no reason why the most fundamental aspects of the laws of nature should be within the grasp of human minds, which evolved for quite different purposes, nor why those laws should have testable consequences at the moderate energies and temperatures that necessarily characterize life-supporting planetary environments. There are further barriers to the questions we may ask of the universe, and the answers that it can provide us with. These are barriers imposed by the nature of knowledge itself, not by human fallibility or technical limitations. As we probe deeper into the intertwined logical structures that underwrite the nature of reality, we can expect to find more of these deep results which limit what

can be known. Ultimately, we may even find that their totality charac-
terizes the universe more precisely than the catalogue of those things that
we can know.

Barrow later expanded these sentiments in his 1998 book *Impossibility:
The Limits of Science and the Science of Limits* (Oxford University Press).
Here are some passages from his final courageous chapter:

> The idea that some things may be unachievable or unimaginable tends
> to produce an explosion of knee-jerk reactions amongst scientific (and
> not so scientific) commentators. Some see it as an affront to the spirit of
> human inquiry: raising the white flag to the forces of ignorance. Others
> fear that talk of the impossible plays into the hands of the anti-scientists,
> airing doubts that should be left unsaid lest they undermine the public
> perception of science as a never-ending success story. . . .
>
> We live in strange times. We also live in strange places. As we probe
> deeper into the intertwined logical structures that underwrite the nature
> of reality, I believe that we can expect to find more of these deep results
> which limit what can be known. Our knowledge about the Universe has
> an edge. Ultimately, we may even find that the fractal edge of our knowl-
> edge of the Universe defines its character more precisely than its contents;
> that what cannot be known is more revealing than what can.

George Gamow once described science as an expanding circle, not on a
plane but on a sphere. It reaches a maximum size, after which it starts to
contract until finally the sphere is covered and no more fundamental knowledge about the universe remains. In recent years numerous physicists,
Hawking for instance, have expressed similar hopes. Richard Feynman
suggested that although the circle may start to contract, it will become ever
more difficult to obtain new knowledge and close the circle completely.

That science will soon discover everything is far from a recent hope.
William James, lecturing at Harvard more than a century ago, attacked the
hope with these words:

In this very University . . . I have heard more than one teacher say that all the fundamental conceptions of truth have already been found by science, and that the future has only the details of the picture to fill in. But the slightest reflection . . . will suffice to show how barbaric such notions are. They show such a lack of scientific imagination, that it is hard to see how one who is actively advancing any part of science can make a mistake so crude. . . .

Our science is a drop, our ignorance a sea. Whatever else be certain, this at least is certain—that the world of our present natural knowledge *is* enveloped in a larger world of *some* sort of whose residual properties we at present can frame no positive idea.

Infinite In All Directions, the title of Freeman Dyson's 1988 book, says it all. Near the close of his third chapter he has this to say about a different hope:

It is my hope that we may be able to prove the world of physics as inexhaustible as the world of mathematics. Some of our colleagues in particle physics think that they are coming close to a complete understanding of the basic laws of nature. They have indeed made wonderful progress in the last ten years. But I hope that the notion of a final statement of the laws of physics will prove as illusory as the notion of a formal decision process for all of mathematics. If it should turn out that the whole of physical reality can be described by a finite set of equations, I would be disappointed. I would feel that the Creator had been uncharacteristically lacking in imagination. I would have to say, as Einstein once said in a similar context, *"Da könnt' mir halt der liebe Gott leid tun"* ("Then I would have been sorry for the dear Lord").

Addendum

British philosopher Derek Parfit, in a two-part essay titled "Why Anything? Why This?" in the *London Review of Books* (January 22 and February 5,

1998), struggles brilliantly with the superultimate question of why there is something. That nothing should exist is much simpler than the reality of a universe in which, as he says, even the brain of an earthworm is more complicated than a lifeless galaxy.

How can we explain the brute fact that the universe, including such bizarre creatures as you and me, manages to exist? Many physicists now think the big bang began as a random fluctuation in a vacuum. "But what physicists call a vacuum isn't really nothing. We can ask why it exists, and has the potentialities it does. In Hawking's phrase, 'What breathes fire into the equations?' "

"Consider first the Null Possibility," Parfit continues, "in which nothing ever exists. To imagine this possibility, it might help to suppose first, that all that ever existed was a single atom. We then imagine that even this atom never existed."

After a few minutes of painful meditation on the Null Possibility, the fact that anything exists, Parfit writes with understatement, "takes one's breath away." It is not clear whether he thinks one is justified in hanging the universe on God, or whether he thinks such an emotional leap has no value. In any case, Parfit obviously cannot answer the ultimate why.

Philosopher Nicholas Rescher, in his book *The Limits of Science* (1984), provides what he claims is a formal proof that it is logically impossible for science ever to know everything. Albert Michelson is often ridiculed for his remark that all the great discoveries in physics have been made, and now it is only a matter of adding more decimal places. As Robert Millikan points out in his autobiography, Michelson attributed his remark to an "eminent physicist," probably Lord Kelvin, but later, "in conversation with me, he was to upbraid himself roundly for this remark."

What remains to Be Discovered (1998), by Sir John Maddox, the distinguished former editor of *Nature,* is a splendid survey of the great unanswered questions facing today's scientists. He agrees with those who believe that science is a never-ending quest; that every question answered raises another, and that there will be no end to the process.

Hugo Gernsback headed his monthly editorial in the November 1926 issue of *Science and Invention,* his marvelous forgotten magazine, "Can We

Reach the End of Knowledge?" Like Maddox, he surveys the big unknown questions facing the science of his day and concludes that what science knows is infinitesimal compared to what will become known in the next ten thousand years.

"As yet we have really seen nothing," Gernsback writes. "We are still groping in the dark and our minds are still extremely primitive. . . . We still look at all the really big things as a dog might look at a complicated radio set. The dog can see the set and hear the sounds. He knows that it is a physical object but that is about all. Our present knowledge is very similar to the dog's knowledge."

At the end of 100 million years, Gernsback asks, if the human race survives that long, will the end of knowledge be reached? He thinks not. "The human race will be just as far from the end of knowledge as we are now."

Index

About the Author

Born in 1914, Martin Gardner obtained a bachelor's degree in philosophy at the University of Chicago in 1936. Early jobs included reporter on the *Tulsa Tribune,* writer in the University of Chicago's press relations office, and case worker in Chicago's Black Belt. After four years as a Navy yeoman he began freelancing with sales of fiction to *Esquire.* In New York City he worked eight years as contributing editor of *Humpty Dumpty's Magazine* before he began a twenty-five year period with *Scientific American* as writer of the Mathematical Games column.

Gardner is the author of more than seventy books about mathematics, science, philosophy, and literature. He has written two novels, *The Flight of Peter Fromm* and *Visitors from Oz*, and a collection of short stories, *The No-Sided Professor*. His two *Annotated Alice* books have been combined in a single volume published in 1999 by W. W. Norton. Among his awards are two honorary doctorates and several prizes for science and math writing. His main hobby is magic, about which he has written several technical books.

Gardner calls himself a philosophical theist in the tradition of Plato, Kant, Pierre Bayle, Charles Peirce, William James, and Miguel de Unamuno. He and his wife, Charlotte, live quietly in the mountains of western North Carolina.